MATHEMATICAL FOUNDATIONS OF QUANTUM MECHANICS

Mathematical Foundations of Quantum Mechanics

George W. Mackey
Professor of Mathematics
Harvard University

DOVER PUBLICATIONS, INC.
Mineola, New York

Bibliographical Note

This Dover edition, first published in 2004, is an unabridged republication of the work originally published in the "Mathematical Physics Monograph Series" by W. A. Benjamin, Inc., New York, in 1963.

Library of Congress Cataloging-in-Publication Data

Mackey, George Whitelaw, 1916–
Mathematical foundations of quantum mechanics / George W. Mackey.
p. cm.
Originally published: New York : W.A. Benjamin, 1963 in series: Mathematical physics monograph series.
Includes bibliographical references.
ISBN 0-486-43517-2 (pbk.)
1. Quantum theory. 2. Mathematical physics. I. Title.

QC174.12.M28 2004
530.12—dc22

2003063503

Manufactured in the United States of America
Dover Publications, Inc., 31 East 2nd Street, Mineola, N.Y. 11501

EDITOR'S FOREWORD

It is a truism based on several thousand years experience that the interaction between mathematics and physics can be fruitful for both. The subtlety and depth of modern physical theories makes essential systematic studies of their logical and mathematical structure, yet such investigations have been rather rarely available in the books published in recent years. The purpose of the present series is to provide a home for such books. Mathematical physics is herewith defined as the pursuit of significant structure in physical theory. It is hoped that this series will make readily available systematic accounts of recent developments in mathematical physics.

A. S. Wightman

Princeton, New Jersey
August 1963

PREFACE

When lecturing on advanced topics the author frequently writes out a more or less complete (and somewhat improved) draft of the lectures actually given and makes them available to the students. This was done in particular for a course in the mathematical foundations of quantum mechanics given at Harvard in the spring of 1960. These notes were corrected, typed, and mimeographed by Messrs. E. Bolker, V. Manjarrez, A. Ramsay, and M. Spivak and put on sale by the Harvard Mathematics Department. The text of this book is substantially that of those notes. However, several pages have been radically revised, numerous small errors have been corrected, and a short appendix has been added.

The course (Mathematics 263) for which the notes were written was designed for students with a reasonably high degree of facility in dealing with abstract mathematical concepts and little or no knowledge of physics. The reader is assumed to be familiar with the basic concepts of abstract algebra, point set topology, and measure theory, and is given a rapid introduction to coordinate-free tensor analysis on C_∞ manifolds and to the theory of self-adjoint operators in Hilbert space.

The aim of the course was to explain quantum mechanics and certain parts of classical physics from a point of view more congenial to pure mathematicians than that commonly encountered in physics texts. Accordingly, the emphasis is on generality and careful formulation rather than on the technique of solving problems. On the other hand, no attempt is made at complete rigor. In places a complete treatment would have taken us too far afield and in others non-trivial mathematical problems remain to be solved. There are also places where completeness simply seemed more troublesome than illuminating. In sum, we have tried to present an outline of a completely rigorous treatment which can be filled in by any competent mathematician modulo the solution of certain more or less well-defined mathematical problems.

In accordance with our wish to demand no significant physical prerequisites, we have made an attempt to define all physical concepts used in terms of those of pure mathematics and the basic ones of space and time. Our fundamental viewpoint is that the change in time of a physical system may be described by a one-parameter semi-group U acting on a set S and that the laws of physics make assertions about the structure of S and the "infinitesimal generator" of U. In Chapter 1 this viewpoint is developed systematically in so far as it applies to classical mechanics–the various sections dealing with special cases of varying generality. The final section on statistical mechanics forms a natural bridge to quantum mechanics in two different ways. In Chapter 2 quantum mechanics is presented from the same point of view, in such a way as to stress the many parallels with classical mechanics. In particular the last three sections of Chapter 2 correspond one-to-one in a natural way to the last three sections of Chapter 1. Chapter 3, on applications to atomic physics, is much shorter than the others and is somewhat closer in spirit to the conventional treatment.

We have taken advantage of the informal character of a set of notes to indulge in a certain amount of carelessness. If the reader thinks a sign should be changed he is probably right. Perhaps there are more serious errors here and there. We have also been rather haphazard about giving bibliographical indications. Needless to say we have been strongly influenced by the classic treatises of von Neumann and Weyl. Some supplementary bibliographical material will be found in the appendix.

GEORGE W. MACKEY

Cambridge, Massachusetts
July 1963

CONTENTS

MATHEMATICAL FOUNDATIONS OF QUANTUM MECHANICS

Chapter 1

CLASSICAL MECHANICS

1-1 Preliminaries

Let S denote the set of all "states" of a physical system, where "state" is defined in such a way that the state of the system at a time $t = t_0 > 0$ is uniquely determined by the appropriate physical law and the state at $t = 0$. For example, the state of a system of n interacting mass particles is determined by giving the $3n$ position coordinates and the $3n$ velocity coordinates of the n particles. For each $s \in S$ and each $t > 0$ let $U_t(s)$ denote the state at time t when the state at time 0 is s. Then for each fixed t, U_t is a transformation of S into S. Now $U_{t_1}(U_{t_2}(s))$ is the state t_1 time units after the state was $U_{t_2}(s)$ and $U_{t_2}(s)$ is the state t_2 time units after it was s. Thus $U_{t_1}(U_{t_2}(s))$ is the state $t_1 + t_2$ time units after it was s; that is, $U_{t_1+t_2}(s)$. In other words, for all t_1 and t_2 with $t_1 > 0$, $t_2 > 0$ we have

$$U_{t_1+t_2} = U_{t_1} U_{t_2} \tag{1}$$

It follows in particular that the set of all U_t is a *semi-group* of transformations. A semi-group which has been parameterized by the real numbers so that (1) holds is said to be a *one-parameter semi-group*. Thus the change in time of a physical system is described by a one-parameter semi-group. We shall call it the *dynamical semi-group* of the system.

If each U_t is a one-to-one map of S onto S so that U_t^{-1} exists, we shall write $U_{-t} = U_t^{-1}$ and $U_0 = I$, where I is the identity transformation. Then (1) holds for all real t_1 and t_2 and we have a *one-parameter group*. We shall deal mainly with systems that are *reversible* in the sense that the dynamical semi-group may be expanded to a one-parameter group as indicated above.

When our system is reversible, each s lies on one and only one "orbit," where an orbit is the set of all points $U_t(s)$ for fixed s and variable t. Each orbit is a curve in S. Generally speaking (we shall give precise details in various special cases) S has sufficient extra structure so that it makes sense to discuss the "tangent vectors" to the points of each orbit. In this way the dynamical group assigns a "vector" to each point of S, i.e., a "vector field." This vector field is called the "infinitesimal generator" of the group and in many cases determines the group uniquely. This is of great importance because the physical law is usually much more easily expressed by describing the infinitesimal generator of the group than by describing the group itself. Indeed, physical laws are almost always given in infinitesimal form, and in order to obtain the orbits of the group one has to integrate differential equations.

In the special case in which S is an open subset of Euclidean n-space we may make the above considerations much more definite. (We shall consider more general cases later.) Then each orbit in S is a curve in n-space described by n functions of t: $q_1(t), \ldots, q_n(t)$. Here $q_1(t), \ldots, q_n(t) = U_t(q_1(0), \ldots, q_n(0))$. If the derivatives $q_1'(t), \ldots, q_n'(t)$ all exist, they form the components to the tangent vector to the unique orbit through the point $q_1(t), \ldots, q_n(t)$. We shall say then that U is differentiable. Let us denote the n components of the tangent vector to the orbit through $q_1, \ldots, q_n$ at $q_1, \ldots, q_n$ by $A_1^U(q_1, \ldots, q_n)$, $A_2^U(q_1, \ldots, q_n), \ldots, A_n^U(q_1, \ldots, q_n)$. Then every orbit $t \to q_1(t), \ldots, q_n(t)$ satisfies the system of differential equations

$$\left.\begin{aligned} \frac{dq_1}{dt} &= A_1^U(q_1, \ldots, q_n) \\ \frac{dq_2}{dt} &= A_2^U(q_1, \ldots, q_n) \\ &\cdots \\ \frac{dq_n}{dt} &= A_n^U(q_1, \ldots, q_n) \end{aligned}\right\} \qquad (2)$$

If the A_j^U are differentiable functions of $q_1, \ldots, q_n$, then the standard uniqueness theorems for differential equations tell us that there is at most one curve through a given point satisfying (2). Thus U will be uniquely determined by the A_j^U. When the A_j^U exist and are differentiable we shall say that U is twice differentiable. Thus we have a natural one-to-one correspondence between twice-differentiable one-parameter groups in S and *certain* continuous vector fields in S. We may state the physical law by giving explicitly the functions A_j^U.

We remark that not every differentiable vector field in S is the generator of a one-parameter group. The existence theorems in differential equations provide local solutions only and it is easy to give examples in which no global solution (i.e., no group U_t) exists. Moreover, no simple necessary and sufficient conditions for the existence of global solutions are known. On the other hand, it is clear from the above that the vector field cannot define a reversible physical law unless global solutions do exist.

As we shall see later, in systems with an infinite number of degrees of freedom, the above considerations lead to partial differential equations. In quantum mechanics the states can never be described by a finite number of coordinates—even when the corresponding classical states could be. Thus in quantum mechanics we always have a partial differential equation (or a system of such). It is called Schrödinger's equation.

Though one can seldom write it down explicitly, the basic group $t \to U_t$ plays a very important role in theoretical considerations.

1–2 The Laws of Particle Mechanics

Let $x_1, y_1, z_1, x_2, y_2, z_2, \ldots, x_n, y_n, z_n$ be the coordinates of n "particles" in some Euclidean coordinate system. Perhaps the most basic law of classical particle mechanics is that the "future" coordinates are determined by the coordinates and their time derivatives at some particular time. Thus the space S of all states may be identified with a subset of $6n$-dimensional Euclidean space. For the time being we shall suppose that this subset is *open*—which means, roughly speaking, that there are no "constraints." It will be convenient to relabel the coordinates $q_1, \ldots, q_{3n}$ and to denote the corresponding time derivatives by $v_1, \ldots, v_{3n}$. Assuming it to be twice-differentiable the dynamical group U can be obtained by integrating a system of ordinary differential equations of the form

$$\frac{dq_j}{dt} = A_j^0(q_1, \ldots, q_{3n}, v_1, \ldots, v_{3n})$$

$$\frac{dv_j}{dt} = A_j(q_1, \ldots, q_{3n}, v_1, \ldots, v_{3n})$$

Moreover, since $v_j = dq_j/dt$ by definition, the functions A_j^0 are all known and we have the system

$$\frac{dq_j}{dt} = v_j$$

$$\frac{dv_j}{dt} = A_j(q_1, \ldots, q_{3n}, v_1, \ldots, v_{3n})$$

We remark that this is just the system of $6n$ first-order equations obtained from the system of $3n$ second-order equations

$$\frac{d^2q_j}{dt^2} = A_j\left(q_1, \ldots, q_{3n}, \frac{dq_1}{dt}, \ldots, \frac{dq_{3n}}{dt}\right)$$

by the standard device of substituting auxiliary variables for the first derivatives. Further assumptions about the physical laws will restrict the nature of the functions A_j. We shall consider only systems in which the following assumptions are made:

I. The A_j are functions of the q_k alone and are independent of the v_k.

II. There exist positive constants M_j such that

$$\frac{\partial M_j A_j}{\partial q_k} = \frac{\partial M_k A_k}{\partial q_j}$$

III. The $M_j A_j$ are the partial derivatives of a function $-V$.

It is clear that the M_j in II are not uniquely determined by the A_j. We can multiply them all by the same positive constant without altering the truth of II. On the other hand, the ratios M_j/M_k are determined unless the corresponding partial derivatives vanish. If we agree to set $M_j/M_k = 1$ whenever II does not determine some other value for the ratio, we see at once that the M_j are uniquely determined once one of them has been assigned a definite value. Choosing one such value is called "choosing a unit of mass," and the resulting numbers M_j are called the masses associated with the corresponding coordinates. It turns out in practice that $M_{3k+1} = M_{3k+2} = M_{3k+3}$, so that in fact the masses are attributes of the particles. Assumption III is almost a consequence of II. By a well-known result in advanced calculus, V certainly exists locally and these local V's can be combined to form one global one whenever S is simply connected. However, if S is not assumed to be simply connected we must assume III separately.

The function $M_i A_i(q_1, \ldots, q_{3n})$ is often denoted by $F_i(q_1, \ldots, q_{3n})$ and is called the *force* component acting on the ith coordinate. The number $M_i v_i$ is called the *momentum* component conjugate to q_i. In terms of the forces and momenta the equations of motion take the form

$$\frac{dq_i}{dt} = \frac{p_i}{M_i}, \quad \frac{dp_i}{dt} = F_i(q_1, \ldots, q_{3n})$$

Assumption III takes the form $F_i = -\partial V/\partial q_i$. One says that the forces are *conservative* and are derived from the *potential* V.

Since the v_i and p_i determine one another uniquely we may regard the state of our system as described by the q_i and the p_i instead of by the q_i and v_i. Of course, when this is done S becomes a different subset of 6n-dimensional space. When S is the set of all possible q_i and p_i it is called phase space. The real significance of the switch to phase space will become clearer in the coordinate-free treatment to be given later.

By an *integral* of our system we shall mean a function ϕ defined on phase space S such that ϕ is constant on the U_t orbits. If ϕ is differentiable, then $d/dt[\phi(U_t(s))]_{t=0}$ is easily seen to be

$$\frac{\partial\phi}{\partial q_1}\frac{dq_1}{dt} + \cdots + \frac{\partial\phi}{\partial q_{3n}}\frac{dq_{3n}}{dt} + \frac{\partial\phi}{\partial p_1}\frac{dp_1}{dt} + \cdots + \frac{\partial\phi}{\partial p_{3n}}\frac{dp_{3n}}{dt}$$

$$= \frac{\partial\phi}{\partial q_1}\frac{p_1}{M_1} + \cdots + \frac{\partial\phi}{\partial q_{3n}}\frac{p_{3n}}{M_{3n}} - \frac{\partial\phi}{\partial p_1}\frac{\partial V}{\partial q_1} - \cdots - \frac{\partial\phi}{\partial p_{3n}}\frac{\partial V}{\partial q_{3n}}$$

Thus ϕ is an integral if and only if this last expression is identically zero in the q's and p's.

More generally, let W be any twice-differentiable one-parameter group and let the 6n components of its infinitesimal generator be denoted by $B^W_1, \ldots, B^W_{3n}, C^W_1, \ldots, C^W_{3n}$, each B^W_i and C^W_j being a function of the q's and p's. Then a function ϕ is constant on the orbits of W if and only if

$$0 = \frac{\partial\phi}{\partial q_1} B^W_1 + \cdots + \frac{\partial\phi}{\partial q_{3n}} B^W_{3n} + \frac{\partial\phi}{\partial p_1} C^W_1 + \cdots + \frac{\partial\phi}{\partial p_{3n}} C^W_{3n}$$

Suppose that the vector field whose components, in order, are the C^W_j and $-B^W_j$ is the set of all partial derivatives of some function ϕ; that is, suppose that

$$C^W_j = \frac{\partial\phi}{\partial q_j} \qquad B^W_j = -\frac{\partial\phi}{\partial p_j}$$

It follows at once from the above identity that ϕ will be constant on the W orbits. Such vector fields play an important role in the theory. They are called *infinitesimal contact transformations*. If the infinitesimal generator of W is an infinitesimal contact transformation, that is, if ϕ exists so that

$$C^W_j = \frac{\partial\phi}{\partial q_j} \qquad B^W_j = -\frac{\partial\phi}{\partial p_j}$$

we say that W is a *one-parameter group of contact transformations*.

The function ϕ determines W uniquely and is uniquely determined by it up to an additive constant. We shall call it the *fundamental invariant* of W.

We shall now show that our dynamical group U is a one-parameter group of contact transformations and hence has at least one nontrivial integral. We must find a function (which we shall call H) such that

$$\frac{\partial H}{\partial p_j} = \frac{p_j}{M_j} \quad \text{and} \quad \frac{\partial H}{\partial q_j} = \frac{\partial V}{\partial q_j}$$

We see from the first set of equations that H must be of the form

$$\frac{p_1^2}{2M_1} + \frac{p_2^2}{2M_2} + \cdots + \frac{p_{3n}^2}{2M_{3n}} + H_0(q_1, \ldots, q_{3n})$$

and from the second that we may take $H_0 = V$. Thus the function H, where

$$H(q_1, \ldots, q_{3n}, p_1, \ldots, p_{3n}) = \frac{p_1^2}{2M_1} + \cdots + \frac{p_{3n}^2}{2M_{3n}} + V(q_1, \ldots, q_{3n})$$

is a constant on the orbits of U. It is called the *integral of energy* or simply the *energy* of the system. The fact that it remains constant in time is one aspect of the so-called "law of conservation of energy." In terms of H we may rewrite the differential equations of motion in so-called "Hamiltonian" form,

$$\frac{dq_i}{dt} = \frac{\partial H}{\partial p_i} \qquad \frac{dp_i}{dt} = -\frac{\partial H}{\partial q_i}$$

In this context the function H is called the Hamiltonian of the system. We note that H is the sum of two terms, one of which depends only upon the positions and the other only upon the velocities. These two terms are known respectively as the *potential energy* and the *kinetic energy*.

Let W be a one-parameter group of contact transformations whose fundamental invariant is ψ and let us consider the condition that ϕ be a constant on the orbits of W. Substituting in the formula derived above we find that the condition is

$$-\frac{\partial \phi}{\partial q_1}\frac{\partial \psi}{\partial p_1} - \cdots - \frac{\partial \phi}{\partial q_{3n}}\frac{\partial \psi}{\partial p_{3n}} + \frac{\partial \phi}{\partial p_1}\frac{\partial \psi}{\partial q_1} + \cdots + \frac{\partial \phi}{\partial p_{3n}}\frac{\partial \psi}{\partial q_{3n}} = 0$$

The expression on the left is called the *Poisson bracket* of ϕ and ψ and is denoted by $[\phi, \psi]$. It is obvious that $[\phi, \psi] = -[\psi, \phi]$ and hence that $[\phi,\psi] \equiv 0$ if and only if $[\psi,\phi] \equiv 0$. This means that ϕ is a constant on the orbits of the one-parameter group of contact transformations defined by ψ if and only if ψ is a constant on the orbits of the one-parameter group of contact transformations defined by ϕ. In the special case in which $\phi = H$ we get the following important principle. Let ψ be the fundamental invariant of any one-parameter group of contact transformations W^ψ. Then ψ is an integral of our dynamical system if and only if the transformations W^ψ carry H into itself. In this way we get a correspondence between integrals and one-parameter groups of "symmetries." As we shall now show, the familiar momentum integrals correspond to the translational and rotational symmetries of space.

Suppose that for each fixed $q_1, \ldots, q_{3n}$ S contains $q_1, \ldots, q_{3n}$, $p_1, \ldots, p_{3n}$ for arbitrary $p_1, \ldots, p_{3n}$. Let $\mathfrak{M}$ denote the open set in E^{3n} consisting of all $q_1, \ldots, q_{3n}$ which occur. Let $t \to U_t$ be a twice-differentiable one-parameter group in $\mathfrak{M}$ whose infinitesimal generator is the vector field with components $D_1, \ldots, D_{3n}$. Then in a manner which it will be easier to explain in the next section, U induces a one-parameter group W in S, whose infinitesimal generator is

$$D_1, \ldots, D_{3n}, -\sum_{i=1}^{3n} p_i \frac{\partial D_i}{\partial q_1}, -\sum_{i=1}^{3n} p_i \frac{\partial D_i}{\partial q_2}, \ldots, -\sum_{i=1}^{3n} p_i \frac{\partial D_i}{\partial q_{3n}}$$

We see at once that W is a one-parameter group of contact transformations whose fundamental invariant is $p_1D_1(q_1, \ldots, q_{3n}) + \cdots + p_{3n}D_{3n}(q_1, \ldots, q_{3n})$. Whenever U is such that H is left invariant by all W_t the function $p_1D_1(q_1, \ldots, q_{3n}) + \cdots + p_{3n}D_{3n}(q_1, \ldots, q_{3n})$ will be an integral which is linear in the p's. Such integrals, when they exist, are called *momentum integrals*.

An important case in which momentum integrals occur is that in which V depends only upon the distances between the particles—for instance, in planetary motion or more generally when

$$F_i(x_1, y_1, z_1, \ldots, z_n) = \sum_{\substack{j=1 \\ j \neq i}}^{n} G_{ij}(|w_i - w_j|) \frac{w_i - w_j}{|w_i - w_j|}$$

where w_i is the vector x_i, y_i, z_i, $|w_i - w_j| = \sqrt{(x_i - x_j)^2 + (y_i - y_j)^2 + (z_i - z_j)^2}$ and G_{ij} is a continuous function defined on the positive real axis such that $G_{ij} = G_{ji}$. In such a case let $t \to A_t$ be any one-parameter group of distance preserving transformations in 3-space. Then $x_1, y_1, z_1, \ldots, x_n, y_n, z_n \to$

$A_t(x_1, y_1, z_1), \ldots, A_t(x_n, y_n, z_n)$ defines a one-parameter group V in $\mathfrak{M}$ such that the corresponding W in S leaves H fixed. This will be immediately obvious from the coordinate-free considerations to be given in the next section and can be verified by suitable computation now. In this way we associate an integral which is linear in the momenta with each group $t \to A_t$. If $A_t(x_1, y_1, z_1) = x_1 + t, y_1, z_1, x_2 + t, y_2, z_2, \ldots,$ this integral turns out to be $p_1 + p_4 + p_7 + \cdots$. Similarly the group of all translations in the y direction leads to the integral $p_2 + p_5 + p_8 + \cdots$ and the group of all translations in the z direction to the integral $p_3 + p_6 + p_9 + \cdots$. These three integrals are called respectively the x, y, and z components of the total (linear) momentum. The fact that they are integrals is the content of the law of conservation of momentum. Translating in other directions one gets other integrals, but these give nothing really new. They are all linear combinations with constant coefficients of the integrals just described.

Now consider the group $x, y, z \to x \cos t + y \sin t, -x \sin t + y \cos t, z$ of all rotations about the z axis. This leads to an integral

$$\sum_{j=1}^{n} (p_{3j-2}\, q_{3j-1} - q_{3j-2}\, p_{3j-1})$$

called the angular momentum about the z axis. Similarly, one is led to define the angular momenta about the x and y axes. Angular momenta about axes in other directions through points other than $0, 0, 0$ are again integrals but are simply expressible in terms of the six already described. We emphasize these group theoretical definitions of the linear and angular momenta because they are the definitions which generalize most naturally to quantum mechanics.

The mapping $q_1, \ldots, q_{3n}, p_1, \ldots, p_{3n} \to q_1, \ldots, q_{3n}, p_1/M_1, \ldots, p_{3n}/M_{3n}$ is one to one from phase space S regarded as the set of possible values of the q's and p's onto the original state space S' of all possible values of the q's and v's. This mapping takes the function H on S into the function H' on S' where

$$H'(q_1, \ldots, q_{3n}, v_1, \ldots, v_{3n}) = \frac{M_1 v_1^2}{2} + \cdots + \frac{M_{3n} v_{3n}^2}{2} + V(q_1, \ldots, q_{3n})$$

Expressed in terms of H' the equations of motion are

$$\frac{dq_i}{dt} = v_i$$

$$\frac{dv_i}{dt} = -\frac{\partial H'}{\partial q_i} \frac{1}{M_i}$$

Now

$$v_i = \frac{\partial H'}{\partial v_i}\frac{1}{M_i}$$

so we may rewrite these as

$$\frac{d}{dt}\left(\frac{\partial H'}{\partial v_i}\right) = -\frac{\partial H'}{\partial q_i}$$

$$\frac{dq_i}{dt} = v_i$$

Finally if we let $L(q_1, \ldots, q_{3n}, v_1, \ldots, v_{3n}) =$

$$\frac{M_1 v_1^2}{2} + \cdots + \frac{M_{3n} v_{3n}^2}{2} - V(q_1, \ldots, q_{3n})$$

we get

$$\frac{d}{dt}\left(\frac{\partial L}{\partial v_i}\right) = \frac{\partial L}{\partial q_i} \qquad \frac{dq_i}{dt} = v_i$$

or in the more elliptical form which is customary,

$$\frac{d}{dt}\left(\frac{\partial L}{\partial \dot{q}_i}\right) = \frac{\partial L}{\partial q_i}$$

So written the equations of motion are said to be in *Lagrangian form* and L is called the Lagrangian of the system. The interest of Lagrange's equations lies in the fact that they are precisely the Euler equations for a certain calculus of variations problem. Thus the orbits of U_t are defined by the function L in an intrinsic manner having nothing to do with the particular coordinate system used. In fact, let L be any suitably differentiable function defined on S' and suppose that for each $q_1, \ldots, q_{3n}$ S' contains $q_1, \ldots, q_{3n}, v_1, \ldots, v_{3n}$ for all possible values of $v_1, \ldots, v_{3n}$. Let $\mathfrak{M}$ be the set of all possible 3n-tuples $q_1, \ldots, q_{3n}$ arising from points of S'. Let q' and q'' be two points in $\mathfrak{M}$ and let $q_1(t), \ldots, q_{3n}(t)$ be differentiable functions on the unit interval such that $q_i(0) = q_i'$ and $q_i(1) = q_i''$. Then the 6n-tuple $q_1(t), \ldots, q_{3n}(t), \dot{q}_1(t), \ldots, \dot{q}_{3n}(t)$ may be substituted in L and integrated with respect to t from 0 to 1. In this way we assign a number $I(C)$ to each differentiable curve C in $\mathfrak{M}$ joining q' to q''. We seek conditions on the $q_i(t)$ ensuring that the corresponding curve gives a "stationary" value to $I(C)$. More precisely, if $\eta_i(t)$ is such that $\eta_i(0) = \eta_i(1) = 0$, then

$q_i(t) + \epsilon \eta_i(t) = C_\epsilon^\eta$ is an admissible curve for each ϵ. We say that C gives a stationary value to I if $(d/d\epsilon) I(C_\epsilon^\eta)_{\epsilon=0} = 0$ for every η. By a reasonably straightforward argument, which the reader will find in any text on the calculus of variations, it can be shown that the curve defined by the q_i gives a stationary value to I if and only if the q_i satisfy the differential equation

$$\frac{d}{dt}\left(\frac{\partial L}{\partial \dot{q}_i}(q_1(t), \ldots, q_{3n}(t), \dot{q}_1(t), \ldots, \dot{q}_{3n}(t))\right)$$

$$\equiv \frac{\partial L}{\partial q_i}(q_1(t), \ldots, q_{3n}(t), \dot{q}_1(t), \ldots, q_{3n}(t))$$

When L is the Lagrangian of our system there are just Lagrange's equations. The fact that the orbits of U_t make the indicated integral stationary is known as *Hamilton's principle.*

1-3 Generalized Coordinates and Differentiable Manifolds

It is sometimes convenient to switch to a new coordinate system in which the coordinates are not only not rectangular but may depend upon the rectangular coordinates of several different particles. Moreover in dealing with a system in which there are "constraints" it is convenient to reduce the number of coordinates to the number which can be independently varied and this may be only possible locally. For example, if $\mathfrak{M}$ is the surface of a sphere it is impossible to find two continuous functions f and g such that $q \to f(q), g(q)$ is one-to-one from $\mathfrak{M}$ onto a subset of the plane. Finally one often obtains deeper insights into the nature of things by formulating results in a coordinate-free manner. For these reasons we find it worthwhile to give the following rather abstract generalization of the material covered in Secs. 1-1 and 1-2.

To motivate our later definition we begin by proving a theorem characterizing vectors in E^n by the partial derivatives which they define. Let $\mathcal{O}$ be an open subset of E^n and let $\mathcal{E}_\mathcal{O}$ denote the set of all real valued functions which are defined in $\mathcal{O}$ and are of class C_∞ in the sense that they have continuous partial derivatives of all orders. $\mathcal{E}_\mathcal{O}$ is a *ring* of functions in the sense that it is closed under both addition and multiplication. For each $f \in \mathcal{E}_\mathcal{O}$, each $a = a_1, \ldots, a_n \in \mathcal{O}$ and each $v = v_1, \ldots, v_n \in E^n$ consider the real valued function on the line $t \to f(a + tv)$. It is defined for all sufficiently small t and is also in C_∞. We define $f_v(a)$ as

$$\frac{d}{dt} f(a + tv)\bigg]_{t=0} = \lim_{t \to 0} \frac{f(a + tv) - f(a)}{t}$$

We note that when $v = 1, 0, \ldots, 0$ then $f_v(a) = \partial f/\partial x_1(a_1, \ldots, a_n)$ and similarly for the other partial derivatives. More generally,

$$f_v(a) = v_1 \frac{\partial f}{\partial x_1}(a_1, \ldots, a_n) + v_2 \frac{\partial f}{\partial x_2}(a_1, \ldots, a_n) + \cdots$$

$$+ v_n \frac{\partial f}{\partial x_n}(a_1, \ldots, a_n)$$

Thus $f_v(a)$ is always a finite linear combination of the first partial derivatives of f at a. Since $f_{\lambda v}(a) = \lambda f_v(a)$ we see that $f_v(a)$ has the following geometric interpretation. It is the length of v multiplied by the directional derivative of f in the direction of v. For fixed a and v consider the functional $\ell^{a,v}$ taking the function $f \in \mathcal{E}_{\mathcal{O}}$ into the number $f_v(a)$. As is readily verified, $\ell^{a,v}$ has the following formal properties:

(a) $\ell^{a,v}(f + g) = \ell^{a,v}(f) + \ell^{a,v}(g)$
(b) $\ell^{a,v}(fg) = f(a)\,\ell^{a,v}(g) + g(a)\,\ell^{a,v}(f)$
(c) $\ell^{a,v}(f) = 0$ whenever f is a constant

Moreover, if $\ell^{a,v}(f) = \ell^{a,w}(f)$ for all f, the $v = w$. Thus the mapping $v \to \ell^{a,v}$ is one-to-one; that is, the vector v is completely described by the functional $\ell^{a,v}$. We show now that every functional with the properties (a), (b), and (c) is associated with a vector.

Theorem: Let a be a point of $\mathcal{O}$ and let ℓ be any functional taking each function $f \in \mathcal{E}_{\mathcal{O}}$ into a number $\ell(f)$ in such a manner that

I: $\ell(f + g) = \ell(f) + \ell(g)$
II: $\ell(fg) = f(a)\,\ell(g) + g(a)\,\ell(f)$
(I and II: for all f and g in $\mathcal{E}_{\mathcal{O}}$)
III: $\ell(f) = 0$ whenever f is constant

Then there exists a unique vector v such that $\ell(f) = f_v(a)$ for all $f \in \mathcal{E}_{\mathcal{O}}$.

Proof: We note first that II and III together imply that $\ell(\lambda f) = \lambda\,\ell(f)$ whenever λ is a constant. Suppose now that $\mathcal{O}$ is not only open but is convex as well. Given $f \in \mathcal{E}_{\mathcal{O}}$ and $x \in \mathcal{O}$, let $\phi(t) = f(a + t(x - a))$. Then

$$f(x) - f(a) = \phi(1) - \phi(0) = \int_0^1 \phi'(t)\,dt$$

$$= \int_0^1 \sum_{j=1}^{n} (x_j - a_j)\, f_j(a + t(x - a))$$

where f_j denotes the first partial derivative of f with respect to the j-th variable. Let

$$A_j(x_1, \ldots, x_n) = \int_0^1 f_j(a + t(x - a))\,dt$$

Then A_j is easily seen to be in $\mathcal{E}_{\mathcal{O}}$ and we have the identity

$$f(x_1, \ldots, x_n) = f(a_1, \ldots, a_n) + \sum_{j=1}^{n} (x_j - a_j) A_j(x_1, \ldots, x_n)$$

Applying the same argument to each A_j we get

$$A_j(x_1, \ldots, x_n) = c_j + \sum_{i=1}^{n} (x_i - a_i) A_{ij}(x_1, \ldots, x_n)$$

where $c_j = A_j(a) = f_j(a)$. Substituting we obtain the new identity $f(x_1, \ldots, x_n) \equiv f(a_1, \ldots, a_n) + \sum c_j(x_j - a_j) + \sum (x_i - a_i)(x_j - a_j) A_{ij}(x_1, \ldots, x_n)$. Thus $\ell(f) = 0 + \sum c_j \ell(x_j) + 0$. The last term is zero because it is a sum of terms each of which has *two* factors which vanish at a. [$\ell(h_1 h_2 h_3) = h_1(a)\ell(h_2 h_3) + = h_2(a)h_3(a)\ell(h_1) = 0$ if $h_1(a) = h_2(a) = 0$.] Let $v_j = \ell(x_j)$. Then $\ell(f) = \sum v_j f_j(a) = \ell^{a,v}(f)$.

Now suppose that $\mathcal{O}$ is not necessarily convex. We first remark that if f is 0 in an open set about a then $\ell(f) = 0$. In fact we can always find $g \in \mathcal{E}_{\mathcal{O}}$ such that $g(a) = 1$ and $g(x)$ is 0 outside of the set where f is 0. Thus $fg = 0$, so $\ell(fg) = f(a)\ell(g) + g(a)\ell(f) = 0 + \ell(f) = \ell(f) = 0$. It follows that if f_1 and f_2 agree in a neighborhood of a, then $\ell(f_1) = \ell(f_2)$. Now choose an open set $\mathcal{O}'$ such that $\mathcal{O}'$ is convex and $a \in \mathcal{O}' \subseteq \mathcal{O}$. For each member f of $\mathcal{E}_{\mathcal{O}'_v}$ choose $g \in \mathcal{E}_{\mathcal{O}}$ so that $f(x) = g(x)$ in some neighborhood of a. If g_1 and g_2 are two such, then by the above $\ell(g_1) = \ell(g_2)$. Hence $\ell(g)$ is independent of the choice of g and depends only on f. Call it $\bar{\ell}(f)$. $\bar{\ell}$ is easily seen to have properties I, II, III, and since $\mathcal{O}$ is convex there exists a vector v such that $\bar{\ell}(f) = \ell^{a,v}(f)$ for all $f \in \mathcal{E}_{\mathcal{O}'}$. But if $f \in \mathcal{E}_{\mathcal{O}}$ then $\ell(f) = \bar{\ell}(g)$ for some g in $\mathcal{E}_{\mathcal{O}'}$, which agrees with f in a neighborhood of a. Therefore $\ell(f) = \ell^{a,v}(g) = \ell^{a,v}(f)$ and the proof is complete.

Now let S be any set and let $\mathcal{E}$ be a ring of real-valued functions defined throughout S such that whenever $p \neq q$ there exists f, in $\mathcal{E}$ such that $f(p) \neq f(q)$. S becomes a Hausdorff† space (in fact a

---

†A subset $\mathcal{O}$ of Euclidean space is said to be open if given any point P in $\mathcal{O}$ there exists a positive real number ϵ such that the point P' is in $\mathcal{O}$ whenever the distance between P and P' is less than ϵ. One verifies easily that the open sets have the following properties:

1. The union of any collection of open sets is open.
2. The common part of any two open sets is open.
3. If P and P' are distinct points, then there exist disjoint open sets $\mathcal{O}_1$ and $\mathcal{O}_2$ such that P is in $\mathcal{O}_1$ and P' is in $\mathcal{O}_2$. *(cont'd)*

completely regular space) if we define a set $\mathcal{O} \subseteq S$ to be open whenever for each $q \in \mathcal{O}$ there exists $f_1, \ldots, f_n \in \mathcal{E}$ and $\epsilon > 0$ such that $|f_i(p) - f_i(q)| < \epsilon$ for $i = 1, 2, \ldots, n$ implies $p \in \mathcal{O}$. In this topology all members of $\mathcal{E}$ are continuous and the topology may be equivalently defined as the weakest possible one having this property. Let us say that a real-valued function f is *locally* in $\mathcal{E}$ if each point q at which f is defined is contained in an open set $\mathcal{O}$ such that f is defined throughout $\mathcal{O}$ and coincides there with a member of $\mathcal{E}$. We shall say that $\mathcal{E}$ converts S into a C_∞ *manifold* if it has the following two additional properties:

(a) Any function defined on all of S and locally in $\mathcal{E}$ is actually in $\mathcal{E}$.

(b) For each $q \in S$ there exists an open set $\mathcal{O}$ with $q \in \mathcal{O} \subseteq S$ and a one-to-one bicontinuous mapping θ of some open subset G of Euclidean n-dimensional space E^n onto $\mathcal{O}$ such that f defined in $\mathcal{O}$ is locally in $\mathcal{E}$ if and only if $f \circ \theta$ is an infinitely differentiable function on G.

Let $x_1, \ldots, x_n$ be functions defined in the open subset $\mathcal{O}$ of S and let ϕ denote the mapping $p \to x_1(p), \ldots, x_n(p)$. If ϕ is one-to-one from $\mathcal{O}$ onto some open subset G of E^n and ϕ^{-1} has the properties described in (b), we shall say that $x_1, \ldots, x_n$ form a *coordinate system* for $\mathcal{O}$ and a *local coordinate system* for each point q of $\mathcal{O}$. Clearly the functions in a coordinate system are locally in $\mathcal{E}$. Moreover it follows from (b) that every point q has a local coordinate system, i.e., lies in an open set having a coordinate system. It is not difficult to prove that every point has a local coordinate system consisting entirely of restrictions to the relevant open set of members of $\mathcal{E}$. Of course the members of $\mathcal{E}$ will not usually form a coordinate system for the whole of S.

One may think of a C_∞ manifold as the result of blending together a number of open subsets of Euclidean space in such a manner that the notion of infinitely differentiable function (= C_∞ function) is preserved. We shall in future refer to the members of $\mathcal{E}$ as the C_∞ functions on S and to the functions locally in $\mathcal{E}$ as the functions that are C_∞ in their domains.

If we suppose S to be connected then, as is easy to see, each point will have a local coordinate system with the same number of coordinates as any other. We can then refer to this number as the *dimension* of S. It will be convenient to make this assumption, and from now on it will be understood to have been made.

(ftnt cont'd)

The notion of Hausdorff space results from postulating these properties as axioms. Let S be any set and let $\mathfrak{F}$ be a family of subsets of S. Let us say that a subset of S is open if it is in $\mathfrak{F}$. If the open sets have properties 1, 2, and 3 listed above, we say that S is a Hausdorff space (with respect to the family $\mathfrak{F}$).

Let S be any C_∞ manifold and let $\mathcal{E}$ be the ring of all C_∞ functions on S. By a *vector* at the point q of S we shall mean a real-valued function v on $\mathcal{E}$ such that

(a) $v(f) + v(g) = v(f + g)$

(b) $v(fg) = g(q)v(f) + f(q)v(g)$

(c) $v(c) = 0$

whenever f and g are in $\mathcal{E}$ and c is a constant. We shall often denote $v(f)$ by $f_v(q)$ and call it the "partial derivative" at q defined by v. It is clear that the set of all tangent vectors at a point q forms a vector space over the real numbers. We shall call this the "tangent space" at q and denote it by V_q. Let $x_1, \ldots, x_n$ be a coordinate system for an open set $\mathcal{O}$ containing q. Then for each $f \in \mathcal{E}$ there exists a C_∞ function $\tilde{f}$ defined on an open neighborhood of $x_1(q), \ldots, x_n(q)$ such that $f(p) = \tilde{f}(x_1(p), \ldots, x_n(p))$ for all p in $\mathcal{O}$. We may form $(\partial\tilde{f}/\partial x_j)(x_1(q), \ldots, x_n(q))$ and we shall call these numbers $(\partial/\partial x_1)(f)_q, \ldots, (\partial/\partial x_n)(f)_q$. It follows at once from the theorem proved above that for each j, $f \to (\partial/\partial x_j)(f)_q$ is in V_q and that these n members of V_q form a basis for V_q. Thus we see that in an n-dimensional C_∞ manifold each V_q is an n-dimensional vector space and that each local coordinate system at q defines a basis in V^q. In more down-to-earth language, once we have introduced a coordinate system into the open set $\mathcal{O}$, we may identify it with an open subset of E^n and then the most general tangent vector at the point $q = q_1, \ldots, q_n$ is

$$f \to c_1 \left.\frac{\partial f}{\partial x_1}\right]_q, \ldots, c_n \left.\frac{\partial f}{\partial x_n}\right]_q$$

By a *contravariant vector field* L on the C_∞ manifold S we shall mean an assignment $q \to L_q$ of a member of V_q to each $q \in S$. If L is a contravariant vector field and f is a C_∞ function, then we define a new function $L(f)$, by letting L_q operate on f for each q. In other words, $L(f)(q) \equiv L_q(f)$. If $L(f)$ is a C_∞ function for all f we say that L is a *C_∞ contravariant vector field*. Let $\mathcal{O}$ be an open set in which there is a coordinate system $x_1, \ldots, x_n$. Then each contravariant vector field L is described in $\mathcal{O}$ by its components with respect to the basis in the V_q defined by the coordinates. We have

$$L = A_1(x_1, \ldots, x_n)\frac{\partial}{\partial x_1} + \cdots + A_n(x_1, \ldots, x_n)\frac{\partial}{\partial x_n}$$

and

$$L(f) = A_1(x_1, \ldots, x_n)\frac{\partial f}{\partial x_1} + \cdots$$

It is easily seen that L will be C_∞ if and only if the A_j are C_∞ functions for all local coordinate systems.

Every C_∞ vector field L maps $\mathcal{E}$ into $\mathcal{E}$ in such a manner that the following identities hold:

(a) $L(f + g) = L(f) + L(g)$

(b) $L(fg) = fL(g) + gL(f)$

(c) $L(c) = 0$

for all f and g in $\mathcal{E}$ and all constants c. In other words, L defines a *derivation* of the ring $\mathcal{E}$. Conversely, let D be any derivation of $\mathcal{E}$, that is, any mapping of $\mathcal{E}$ into $\mathcal{E}$ satisfying (a), (b), and (c). For each $q \in S$ let $L_q(f) = D(f)(q)$. We verify at once that L_q is in V_q and that $q \to L_q$ is a contravariant C_∞ vector field such that $L(f) \equiv D(f)$. Thus a C_∞ contravariant vector field may be regarded alternatively as a certain kind of mapping of $\mathcal{E}$ into $\mathcal{E}$.

If L and M are C_∞ contravariant vector fields and we define $(L + M)_q = L_q + M_q$, then $L + M$ is again such. Similarly we may multiply a C_∞ contravariant vector field L by a C_∞ function f and obtain a contravariant vector field fL. We may also define a "product" of L with M but this is less obvious. Given L and M we set $[L,M](f) = L(M(f)) - M(L(f))$ for all $f \in \mathcal{E}$. It is obvious that $f \to [L,M](f)$ has properties (a) and (c) above, and a routine calculation shows that it has (b) as well. Thus $[L,M]$ is also a C_∞ contravariant vector field. The "multiplication" so defined is distributive with respect to addition and is "anti-commutative" in the sense that $[L,M] = -[M,L]$. It is definitely not associative. Instead it satisfies the so-called Jacobi identity

$$[[L,M], N] + [[N,L], M] + [[M,N], L] = 0$$

Thus the set of C_∞ contravariant vector fields forms what is known as a *Lie algebra*.

For each $q \in S$ let V_q^* denote the dual of V_q, that is, the vector space of all linear functionals on V_q. We remind the reader that, although V_q and V_q^* always have the same dimension and thus are isomorphic, there is no "natural" isomorphism of V_q on V_q^*. It is true that a basis $v_1, \ldots, v_n$ in V_q defines a natural dual basis $\ell_1, \ldots, \ell_n$ in V_q^*, ℓ_j being the unique linear functional such that $\ell_k(v_j) = \delta_j^k$. Moreover, given $v_1, \ldots, v_n$ and hence $\ell_1, \ldots, \ell_n$ we may construct an isomorphism $c_1v_1 + \cdots + c_nv_n \to c_1\ell_1 + \cdots + c_n\ell_n$ of V_q on V_q^*. However, this isomorphism depends in general upon which basis is chosen. It is only when we have an inner product in V_q and stick to orthogonal bases that we get a unique isomorphism. Thus in dealing with general coordinates, as we are now doing, it is necessary to make a careful distinction between members of V_q and members of V_q^*. By a *covariant vector field* W in S we shall mean an assignment $q \to W_q$ of a member of V_q^* to each q in S. If W is a covariant vector field and L is a contravariant vector field, then we may form $W_q(L_q)$ for all q and thus obtain a real-valued function $W(L)$ defined on S. If $W(L) \in \mathcal{E}$ whenever L is a C_∞ contravariant vector field we shall say that W is a C_∞ covariant vector field.

Let $f \in \mathcal{E}$. Then for each $q \in S$ and each $v \in V_q$ we may form $f_v(q)$. For fixed q and variable v this is a linear functional. That is, there exists a unique member ℓ_q of V_q^* such that $f_v(q) = \ell_q(v)$ for all $v \in V_q$. This linear functional is called the *differential of f at q* and denoted by $(df)_q$. The C_∞ covariant vector field obtained by assigning $(df)_q$ to q will be denoted simply by df and called the *differential* of f. The fact that df is C_∞ follows at once from the obvious identity $df(L) = L(f)$.

Suppose that $x_1, \ldots, x_n$ is a coordinate system for some open set $\mathcal{O}$ of S. Then at each q in $\mathcal{O}$ $(dx_1)_q, \ldots, (dx_n)_q$ is a basis for V_q^* and, indeed, is the dual basis to the natural basis defined by the coordinate system. This is so because $\partial x_i / \partial x_j = \delta_i^j$. The components of df with respect to this natural basis are at once seen to be $\partial f/\partial x_1, \ldots, \partial f/\partial x_n$. Thus the (covariant) vector field df is just what the physicists refer to as grad f. Since $dx_1, \ldots, dx_n$ are a basis at every $q \in \mathcal{O}$, the most general C_∞ covariant vector field W coincides in $\mathcal{O}$ with $a_1\, dx_1 + \cdots + a_n\, dx_n$ where $a_1, \ldots, a_n$ are suitable C_∞ functions. Thus the most general C_∞ covariant vector field is just a formal differential when expressed in local coordinates. We see at once from this and the well-known facts of advanced calculus that by no means may every C_∞ covariant vector field be put into the form df. Those that can be are said to be *exact*.

Let V be any finite dimensional, real vector space. By a tensor on V of covariant order n and contravariant order m we shall mean a real-valued function F of $n + m$ arguments, the first n of which lie in V and the last m in V^* such that F is linear in each argument separately. For example, a member of V^* is a tensor of covariant order 1, and a member of V can be regarded as a member of V^{**} and hence a tensor of contravariant order 1. An "inner product" in V is a tensor of covariant order 2 having the two additional properties of symmetry and positive definiteness. Let F be any tensor of covariant order 2. Then for each fixed $v_0 \in V$, $F(v_0, v)$ is in V^* as a function of v. Thus there exists a linear mapping $\widetilde{F}$ of V into V^* such that $F(v_0, v) \equiv \widetilde{F}(v_0)(v)$. Conversely, every linear mapping of V into V^* arises in this way from a tensor of covariant order 2. If $\widetilde{F}$ is one-to-one we shall say that F is non-degenerate. Whenever F is non-degenerate we may form $\widetilde{F}^{-1}$ which comes in an analogous fashion from a tensor of contravariant order 2.

By a *tensor field* T of covariant order n and contravariant order m on the C_∞ manifold S, $\mathcal{E}$ we shall mean an assignment of a tensor T_q to each vector space V_q for each $q \in S$. If $L_1, \ldots, L_n$ are contravariant vector fields and $W_1, \ldots, W_m$ are covariant vector fields, we may define a function $T(L_1, \ldots, L_n, W_1, \ldots, W_m)$ by setting $T(L_1, \ldots, L_n, W_1, \ldots, W_m)(q) = T_q(v_1, \ldots, v_n, \ell_1, \ldots, \ell_m)$, where $v_j = L_{j,q}$ and $\ell_k = W_{k,q}$. If $T(L_1, \ldots, L_n, W_1, \ldots, W_m)$ is a C_∞ function whenever the L_j and W_k are C_∞ vector fields, we shall say that T is a C_∞ tensor field. It is clear that

$$T(fL_1' + gL_1'', L_2, \ldots, L_n, W_1, \ldots, W_m)$$

$$= f\,T(L_1', L_2, \ldots, L_n, W_1, \ldots, W_m) + g\,T(L_1'', L_2, \ldots, L_n, W_1, \ldots, W_m)$$

for all f and g in $\mathcal{E}$ and all L_j and W_k and that the analogous identity holds for each of the other arguments. We express this by saying that T is linear over $\mathcal{E}$ in each argument. Conversely, let T' be any function having $n + m$ arguments and values in $\mathcal{E}$ such that the first n arguments are C_∞ contravariant vector fields and the last m are C_∞ covariant vector fields. We leave it to the reader to show that if T' is linear over $\mathcal{E}$ in each argument then it is derivable as indicated above from a unique C_∞ tensor field T.

If T is a tensor field of covariant order n and contravariant order m and $q_1, \ldots, q_k$ is a coordinate system for an open set $\mathcal{O}$, then $q_1, \ldots, q_k$ defines a basis $\phi_1^q, \ldots, \phi_k^q$ in each V_q and a dual basis $\ell_1^q, \ldots, \ell_k^q$ in each V_q^*. The value of T_q is then determined by its values

$$A_{i_1, \ldots, i_n}^{j_1, \ldots, j_m}(q) = T_q(\phi_{i_1}^q, \ldots, \phi_{i_n}^q, \ell_{j_1}^q, \ldots, \ell_{j_m}^q)$$

at sets of basis elements. Thus T can be described (in $\mathcal{O}$) by giving the values of the "index symbols" $A_{i_1, \ldots, i_n}^{j_1, \ldots, j_m}$ at each q. Tensors are often defined in this way.

We shall be chiefly interested in covariant second-order tensors in which the mapping of V_q on V_q^* defined by T_q is non-singular. Such tensors can be used to turn covariant vector fields into contravariant ones and vice versa. In the particular case in which $T(L,M) = T(M,L)$ and $T(L,L) > 0$ except when $L_q = 0$, T is said to be a *Riemannian metric*.

Let W be any C_∞ covariant vector field in S. For each pair L,M of C_∞ contravariant vector fields we define a C_∞ function $\widetilde{W}(L,M)$ as follows: $\widetilde{W}(L,M) = L(W(M)) - M(W(L)) - W([L,M])$. Obviously $\widetilde{W}(L,M) = -\widetilde{W}(M,L)$ and almost as obviously $\widetilde{W}(L_1 + L_2, M) = \widetilde{W}(L_1, M) + \widetilde{W}(L_2, M)$. We shall show that for all $\theta \in \mathcal{E}$, $\widetilde{W}(\theta L, M) = \theta\widetilde{W}(L,M)$ and hence that W is linear over $\mathcal{E}$ in each variable. Note first that

$$\theta L(W(M)) - M(W(\theta L)) = \theta L(W(M)) - M(\theta W(L))$$
$$= \theta L(W(M)) - M(\theta)W(L) - \theta M(W(L))$$
$$= \theta\{L(W(M)) - M(W(L))\} - M(\theta)W(L)$$

On the other hand,

$$[\theta L,M](f) = \theta L(M(f)) - M(\theta L(f)) = \theta L(M(f)) - M(\theta)L(f)$$
$$- \theta M(L(f)) = \theta[L,M](f) - M(\theta)L(f)$$

Thus $[\theta L,M] = \theta[L,M] - M(\theta)L$. Hence $W([\theta L,M]) = \theta W([L,M]) - M(\theta)W(L)$. Thus $W([L,M])$ and $L(W(M)) - M(W(L))$ both fail to be linear over $\mathcal{E}$ by the same correction term. Since $\widetilde{W}$ is the difference between them, this term cancels and $\widetilde{W}$ is linear over $\mathcal{E}$. It follows now from the theorem quoted above that there exists a unique covariant tensor of the second order, which we denote by dW and call the differential of W, such that $dW(L,M) \equiv \widetilde{W}(L,M) \equiv L(W(M)) - M(W(L)) - W([L,M])$. Clearly dW is anti-symmetric in its two arguments.

Let us compute dW in the special case in which $W = df$. We have

$$dW(L,M) = L(W(M)) - M(W(L)) - W([L,M]) = L(df(M))$$
$$- M(df(L)) - df([L,M]) = L(M(f)) - M(L(f)) - [L,M](f) = 0$$

by definition of $[L,M]$. Thus a necessary condition that $W = df$ for some f is that $dW = 0$. This condition is not in general sufficient but is sufficient "locally." That this is so is most easily seen by computing the coefficients of dW in a coordinate system. If $W = a_1 dq_1 + \cdots + a_n dq_n$ then $dW)_{ij}$ turns out to be $(\partial a_i/\partial q_j) - (\partial a_j/\partial q_i)$. Thus $dW = 0$ if and only if $\partial a_i/\partial q_j = \partial a_j/\partial q_i$, and as is well known this condition implies that $a_i = \partial f/\partial q_i$ in some neighborhood of every point.

In general we say that W is *closed* if $dW = 0$. The set of all closed W's is a vector space having the exact W's as a subspace. The dimension of the quotient space depends only on the topology of S and is called "the first Betti number of S." It is 0 when S is the interior of the unit ball in E^n; it is 1 when S is the interior of an annulus.†

---

†For example, let $W = (x\,dy - y\,dx)/(x^2 + y^2)$ and let the region be the annulus $1 < x^2 + y^2 < 2$. The $W = d \arctan y/x$ in suitable subregions but $\arctan y/x$ cannot be defined in a continuous single-valued manner throughout the entire annulus.

Let $t \to q(t)$ be a C_∞ "curve" in S, that is, a function from a connected open subset I of the real line to S such that $f(q(t))$ is C_∞ in t for each $f \in \mathcal{E}$. For each $t_0 \in I$ the mapping

$$f \to v_{t_0}(f) = \frac{d}{dt} f(q(t))_{t=t_0}$$

is easily seen to satisfy the conditions in the definition of tangent vector at $q(t_0)$. Thus $v_{t_0} \in V_{q(t_0)}$. We call v_{t_0} the *tangent vector to the curve* at $q(t_0)$. We note in passing that if G is a Riemannian metric on S then $\sqrt{G(v_t, v_t)}$ is a length for v_t and we may assign a "length" to any finite segment of our curve by forming $\int_a^b \sqrt{G(v_t, v_t)}\, dt$.

Now let $t \to U_t$ be a one-parameter group of automorphisms of S, $\mathcal{E}$ where an automorphism α of S is a one-to-one map of S on S such that $f \in \mathcal{E}$ if and only if $f \circ \alpha \in \mathcal{E}$. Suppose that all the orbits of S under $t \to U_t$ are C_∞ curves. Then, since each q in S lies in exactly one orbit, we get a unique tangent vector at q by taking the tangent to the orbit through q. The resulting vector field we shall denote by L^U and call the *infinitesimal generator* of U. If L^U is a C_∞ vector field we shall say that U is a C_∞ one-parameter group.

Let $\mathcal{O}$ be an open subset of S which admits a coordinate system $q_1, \ldots, q_n$. Then the parts of the orbits of S under U lying in $\mathcal{O}$ are curves which may be described in concrete form as n-tuples of function of t. Just as in Sec. 1-2 these n-tuples satisfy the system of ordinary differential equations

$$\frac{dq_1}{dt} = A_1^U(q_1, \ldots, q_n)$$

$$\ldots$$

$$\frac{dq_n}{dt} = A_n^U(q_1, \ldots, q_n)$$

where $A_1^U, \ldots, A_n^U$ are the components of L^U with respect to the bases defined by the coordinate system. Thus by the classical uniqueness theorem for differential equations U is uniquely determined by L^U whenever U is a C_∞ one-parameter group.

Let S be a C_∞ manifold representing a set of possible "instantaneous configurations" for a physical system. Let $\phi: t \to \phi(t)$ be a C_∞ curve in S. Then at each point t_0 the vector $v_{\phi(t_0)}$ tangent to ϕ at $\phi(t_0)$ represents the rate of change of the system at $t = t_0$ when traveling along the trajectory ϕ. If $\phi(t) = q_1(t), \ldots, q_k(t)$ in some coordinate system, then in this same coordinate system $v_{\phi(t_0)}$ has the components $\dot{q}_1(t_0), \ldots, \dot{q}_k(t_0)$. Thus, in coordinate-free form, giving the q's and $\dot{q}$'s is simply giving a point q_0 in S *and* a vector in V_{q_0}. In other words, in a physical system where states are determined by positional coordinates and velocities, the state manifold has a special kind of structure. It is the manifold of tangent vectors on another manifold of half the dimension. Moreover, as we shall see, the invariant significance of introducing momenta instead of velocities (p's instead of $\dot{q}$'s) is that we choose a particular mapping of each V_q in V_q^* and describe states by members of the V_q^*. Before actually returning to mechanics, therefore, it will be important for us to discuss the special properties of such manifolds of vectors.

Let S be any C_∞ manifold. Let S_V denote the set of all pairs q,v, where $v \in V_q$. If $q_1, \ldots, q_n$ is a coordinate system for an open subset $\mathcal{O}$ of S, then $\partial/\partial q_1, \ldots, \partial/\partial q_n$ is a basis for each V_q with $q \in \mathcal{O}$. Thus we have a natural way of associating a 2n-tuple of real numbers with each point of $\pi^{-1}(\mathcal{O})$, where π is the mapping of S_V onto V which takes q,v into q. We shall say that a function f defined on S_V is in $\mathcal{E}_V$ if it is a C_∞ function on each $\pi^{-1}(\mathcal{O})$ with respect to the indicated coordinate systems. We leave it to the reader to verify that $\mathcal{E}_V$ converts S_V into a C_∞ manifold. This manifold is often called the "tangent bundle" over S.

If S_1 and S_2 are any two C_∞ manifolds and ϕ is a mapping from S_1 to S_2, we say that ϕ is a C_∞ mapping if $f \circ \phi$ is a C_∞ function on S_1 whenever f is a C_∞ function on S_2. Let ϕ be such and let $q \in S_1$. For each $v \in V_q$ we may form $(f \circ \phi)_v(q)$ for each C_∞ function on S_2. For fixed v and variable f we get a function on the C_∞ functions on S_2 which is readily verified to satisfy the conditions laid down in defining tangent vectors at $\phi(q)$. Thus there exists a unique vector w in $V_{\phi(q)}$ such that $(f \circ \phi)_v(q) \equiv f_w(\phi(q))$. w is easily seen to depend linearly upon v. This linear mapping from V_q to $V_{\phi(q)}$ we shall call the differential $d\phi|_q$ of ϕ at q. If $q_1, \ldots, q_n$ is a coordinate system for some open set $\mathcal{O}$ containing q, and $s_1, \ldots, s_m$ is a coordinate system for some open set $\mathcal{O}'$ containing $\phi(q)$, then ϕ will be described by m functions of n real variables $s_j = f_j(q_1, \ldots, q_n)$. Moreover these coordinate systems induce bases in V_q and $V_{\phi(q)}$. It is straightforward to verify that the matrix of $d\phi|_q$ with respect to these coordinate systems is just $||\partial f_j/\partial q_k||$ evaluated at q.

Now let $S_1 = S_V$, $S_2 = S$, and $\phi = \pi$. Then at each point $s,v \in S_V$, $d\pi_{s,v}$ maps $V_{s,v}$ linearly into (actually onto) V_s. If L is any C_∞ contravariant vector field on S_V we may apply $d\pi_{s,v}$ to $L_{s,v}$ and thus get a member of V_s. For each fixed s then we have a mapping $v \to d\pi_{s,v}(L_{s,v})$ of V_s into V_s. Let us call this mapping T_s^L. Let us call the contravariant vector field L in S_V *special* if this mapping is the identity for all s; that is, if $d\pi_{s,v}(L_{s,v}) = v$ for all s and v. If $q_1, \ldots, q_n$ is a coordinate system for the open subset $\mathcal{O}$ of S and $q_1, \ldots, q_n, w_1, \ldots, w_n$ is the corresponding coordinate system for $\pi^{-1}(\mathcal{O})$, then the system of ordinary differential equations associated with a general L takes the form

$$\frac{dq_j}{dt} = B_j(q_1, \ldots, q_n, w_1, \ldots, w_n)$$

$$\frac{dw_j}{dt} = A_j(q_1, \ldots, q_n, w_1, \ldots, w_n)$$

We leave it to the reader to verify that L is special if and only if $B_j(q_1, \ldots, q_n, w_1, \ldots, w_n) \equiv w_j$ for all j. But the first-order systems of the form

$$\frac{dq_j}{dt} = w_j$$

$$\frac{dw_j}{dt} = A_j(q_1, \ldots, q_n, w_1, \ldots, w_n)$$

are just exactly the first-order systems obtained from the second-order systems

$$\frac{d^2q_j}{dt^2} = A_j(q_1, \ldots, q_n, \frac{dq_1}{dt}, \ldots, \frac{dq_n}{dt})$$

by introducing the auxiliary variables $w_j = dq_j/dt$. Thus the invariantly defined mathematical object corresponding to a C_∞ system of second-order differential equations on S is a special C_∞ contravariant vector field on S_V.

Let $\mathcal{L}$ be any C_∞ function on S_V and let $t \to \theta(t)$ denote a C_∞ curve in S. For each t let $\theta'(t)$ denote the tangent vector to the curve at $\theta(t)$. Then the pair $\theta(t), \theta'(t)$ is in S_V for all t and we may form $\mathcal{L}(\theta(t), \theta'(t))$, getting a function of t, which may then be integrated between fixed limits. In coordinates this integral becomes $\int_a^b \mathcal{L}(q_1(t), \ldots, q_n(t), \dot{q}_1(t), \ldots, \dot{q}_n(t))\, dt$, and we see that S_V forms a natural setting for considering the calculus of variation problem envisaged in the formulation of Hamilton's principle.

The set S_{V^*} of all pairs q,w, where $w \in V_q^*$, can be made into a C_∞ manifold by a procedure strictly analogous to that used for S_V. This manifold we shall call the "cotangent bundle" over S and we shall denote the C_∞ mapping $q,w \to q$, as before, by π. By using the mapping $d\pi$ we can single out a particular C_∞ covariant vector field on S_{V^*} which will play a central role in formulating the laws of mechanics. We define this vector field as follows. For each point $q,w \in S_{V^*}$, $d\pi_{q,w}$ is a linear map of $V_{q,w}$ onto V_q. The adjoint $(d\pi)^*_{q,w}$ is then a linear map of V_q^* into $V^*_{q,w}$. Since $w \in V_q^*$ we may apply this linear map to w, getting an element of $V^*_{q,w}$. Thus we have an assignment of a member of $V^*_{q,w}$ to each $q,w \in S_{V^*}$;

that is, a covariant vector field. We shall call this vector field the *fundamental covariant vector field* of the manifold and denote it by W^0.

Let $q_1, \dots, q_n$ be a coordinate system for the open set $\mathcal{O}$ and let $q_1, \dots, q_n, p_1, \dots, p_n$ be the corresponding coordinate system for $\pi^{-1}(\mathcal{O})$. Then the general covariant vector field on $\pi^{-1}(\mathcal{O})$ in S_{V^*} may be written as $a_1 \, dq_1 + \cdots + a_n \, dq_n + b_1 \, dp_1 + \cdots + b_n \, dp_n$, where the a's and the b's are functions of the p's and q's. Let us compute the a's and b's for W^0. Since $\pi(q_1, \dots, q_n, p_1, \dots, p_n) = q_1, \dots, q_n$, $d\pi$ at every point has the matrix

$$\left\| \begin{matrix} 1, 0, \dots, 0 \\ 0, 1., , , .0 \\ \dots \\ 0, 0, \dots, 1 \end{matrix} \quad \bigcirc \right\|$$

and the matrix of $d\pi^*$ is the transpose of this. Hence $(d\pi_{q,p})^*(p) = p,0$. Thus $a_j = p_j$ and $b_j = 0$. In other words, $W^0 = p_1 dq_1 + \cdots + p_n dq_n$. In particular W^0 is a C_∞ covariant vector field, and we may form the C_∞ anti-symmetric, covariant, second-order tensor field dW^0. It is easily computed that the matrix of dW^0 in the coordinate system described above is

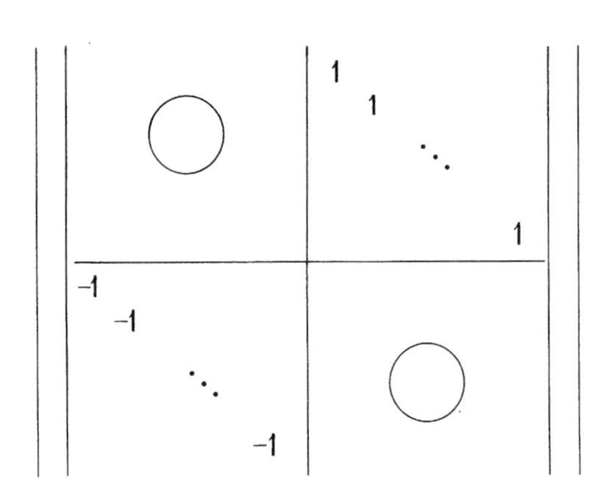

so that dW^0 is everywhere non-singular. For each C_∞ contravariant vector field L let $\widetilde{L}$ be the unique C_∞ covariant vector field such that $\widetilde{L}(M) = dW^0(L,M)$ for all C_∞ contravariant vector fields M. It follows from the non-singularity of dW^0 that $L \to \widetilde{L}$ is one-to-one and onto from the C_∞ contravariant vector fields in S_{V^*} to the C_∞ covariant vector fields in S_{V^*}. We shall use a tilde for the inverse of this mapping as well, so that $\tilde{\tilde{L}} = L$. If

$$L = a_1 \frac{\partial}{\partial q_1} + \cdots + a_n \frac{\partial}{\partial q_n} + b_1 \frac{\partial}{\partial p_1} + \cdots + b_n \frac{\partial}{\partial p_n}$$

then it is easily seen that

$$\tilde{L} = b_1\, dq_1 + \cdots + b_n\, dq_n - a_1\, dp_1 - \cdots - a_n\, dp_n$$

If f and g are C_∞ functions on S_{V^*} then we may form $dW^0(\widetilde{df}, \widetilde{dg})$, which we abbreviate $[f,g]$. In terms of the p's and q's we have

$$[f,g] = \frac{\partial f}{\partial q_1}\frac{\partial g}{\partial p_1} + \cdots + \frac{\partial f}{\partial q_n}\frac{\partial g}{\partial p_n} - \frac{\partial f}{\partial p_1}\frac{\partial g}{\partial q_1} - \cdots - \frac{\partial f}{\partial p_n}\frac{\partial g}{\partial q_n}$$

It is called the "Poisson bracket" of f and g.

The mapping $L \to \tilde{L}$ of contravariant vector fields on covariant vector fields together with the classification of covariant vector fields into those which are closed $(dW = 0)$, those which are exact $(W = dH)$, and those which are neither, produces a corresponding classification of contravariant vector fields. Those of the form $\widetilde{dH}$ we shall call *globally Hamiltonian,* and those of the form $\tilde{W}$, where $dW = 0$, we shall call *locally Hamiltonian.* In the locally Hamiltonian case our vector field will have the form $\widetilde{dH}$ in some neighborhood of each point but there may be no one function H that works for the whole manifold. In any event the differential equation defined by the vector field will take the so-called "Hamiltonian form"

$$\frac{dq_i}{dt} = \frac{\partial H}{\partial p_i} \qquad \frac{dp_i}{dt} = -\frac{\partial H}{\partial q_i}$$

Let U and V denote C_∞ one-parameter groups of automorphisms of S_{V^*} and suppose that their infinitesimal generators L^U and L^V are globally Hamiltonian so that $L^U = \widetilde{df}$ and $L^V = \widetilde{dg}$. Then the rate of change of g along the orbits of U is $L^U(g) = \widetilde{df}(g) = dW^0(\widetilde{df}, \widetilde{dg}) = [f,g]$. Thus the Poisson bracket has the same interpretation as in Sec. 1-2.

The C_∞ one-parameter groups U whose infinitesimal generators L^U are locally Hamiltonian may be neatly characterized as follows. They are just the ones such that each U_t leaves dW^0 invariant. More specifically, an automorphism A of S_{V^*} (as a C_∞ manifold) is said to be a *contact transformation* if $(dA)_{q,w}$ maps $V_{q,w}$ on $V_{A(q,w)}$ in such a manner that $(dW^0)_{q,w}$ is carried into $(dW^0)_{A(q,w)}$. It is then a theorem (which we shall not prove) that the infinitesimal generator of the one-parameter group $t \to U_t$ is locally Hamiltonian if and only if every U_t is a contact transformation. For this reason a locally Hamiltonian contravariant vector field in S_{V^*} is sometimes called an "infinitesimal contact transformation."

We are now ready to return to physics and generalize the laws of particle mechanics as formulated in Sec. 1-2. Instead of assuming that the $3n$ coordinates of our n particles vary in an *open* subset of E^{3n} we assume only that they vary in a subset $\mathfrak{M}$, which becomes a C_∞ manifold when $\mathcal{E}$ is taken to be the set of all restrictions to $\mathfrak{M}$ of C_∞ functions on E^{3n}. For example, we might be given $n(n-1)/2$ constants c_{ij} and let $\mathfrak{M}$ be the set of all $x_1, y_1, z_1, \ldots x_n, y_n, z_n$ such that $(x_i - x_j)^2 + (y_i - y_j)^2 + (z_i - z_j)^2 = c_{ij}$. In order to allow even more general possibilities, for example that space is not "flat" but "curved," we shall, in fact, assume only that we are given an abstract set $\mathfrak{M}$ of "possible configurations" and that $\mathfrak{M}$ is a C_∞ manifold with respect to some set $\mathcal{E}$ of real-valued coordinate functions on $\mathfrak{M}$. If $t \to \phi(t)$ is a curve in $\mathfrak{M}$ representing some possible time sequence of configurations of our system then the tangent vector $\phi'(t)$ will be a measure of the rate at which the system is changing at time t. In fact, if $q_1, \ldots, q_n$ is a local coordinate system so that $t \to \phi(t)$ is described (locally) by a set of n real-valued functions of t, then the derivatives of these functions at t_0 will be the coordinates of $\phi'(t_0)$. Corresponding to our earlier assumption that the future of our system is determined by the coordinates and velocities at one instant of time, we assume now that $\phi(t_0)$ and $\phi'(t_0)$ together determine the future. This means that the states of our system may be identified with points of the tangent bundle $\mathfrak{M}_V$. We shall rule the so-called "non-holonomic" systems out of consideration by assuming that *every* point of $\mathfrak{M}_V$ is a possible state. We shall continue to assume that we have a reversible system and for analytic convenience shall make the unnecessarily strong regularity assumption that the dynamical group $t \to U_t$ is a C_∞ one-parameter group. Of course it follows from the definition of state that the infinitesimal generator L^U of $t \to U_t$ will be a special contravariant vector field in $\mathfrak{M}_V$.

Corresponding to the assumptions I, II, and III made about the acceleration functions A_i in Sec. 1-2, we shall make one assumption that both generalizes and includes all essential features of I, II, and III. Let T denote any C_∞ Riemannian metric in $\mathfrak{M}$. Then at each point q of $\mathfrak{M}$, T_q defines a linear map $\widetilde{T}_q$ of V_q on V_q^*. Using these we may at once define a one-to-one map B^T of $\mathfrak{M}_V$ on $\mathfrak{M}_{V^*}$ by setting $B^T(q,v) = q,\ \widetilde{T}_q(v)$. B^T is easily seen to be C_∞ and to have a C_∞ inverse. Thus if $U_t^T = B^T U_t (B^T)^{-1}$, $t \to U_t^T$ will be a C_∞ one-parameter group of automorphisms of $\mathfrak{M}_{V^*}$. Our fundamental assumption may now be stated as follows:

There exists a C_∞ Riemannian metric T in $\mathfrak{M}$ such that $t \to U_t^T$ has a globally Hamiltonian infinitesimal generator.

Let H be the C_∞ function on $\mathfrak{M}_{V^*}$ such that $\widetilde{dH}$ is the required infinitesimal generator. H is, of course, unique up to an additive constant and is called the Hamiltonian of the system. As we shall see it determines not only U^T but T as well and hence the original dynamical group of our system. Let $q_1, \ldots, q_n$ be a coordinate system for the open subset $\mathcal{O}$ of $\mathfrak{M}$ and let $q_1, \ldots, q_n, v_1, \ldots, v_n$ be the corresponding system for $\pi^{-1}(\mathcal{O})$ in $\mathfrak{M}_V$. Then

$$T_q(v_1, \ldots, v_n, v_1', \ldots, v_n') = \sum_{i,j} g_{ij}(q_1, \ldots, q_n) v_i v_j'$$

so

$$\widetilde{T}_q(v_1, \ldots, v_n) = \sum_{i,j} g_{ij}(q_1, \ldots, q_n) v_j$$

and

$$\widetilde{T}_q^{-1}(p_1, \ldots, p_n) = \sum_{i,j} g_{ij}'(q_1, \ldots, q_n) p_j$$

where the p's are dual to the v's and $\| g_{ij}'(q_1, \ldots, q_n) \| = \| g_{ij}(q_1, \ldots, q_n) \|^{-1}$. We have

$$(B^T)^{-1}(q_1, \ldots, q_n, p_1, \ldots, p_n) = q_1, \ldots, q_n, \sum_j g_{1j}' p_j, \ldots, \sum_j g_{nj}' p_j$$

Thus $dq_i/dt = v_i$ becomes

$$\frac{dq_i}{dt} = \sum_j g_{ij}' p_j$$

Since also $dq_i/dt = \partial H/\partial p_i$, we have

$$\frac{\partial H}{\partial p_i} = \sum g_{ij}' p_j$$

whence

$$H = \frac{\sum g_{ij}' p_i p_j}{2} + \mathcal{V}$$

where $\mathcal{V}$ is a function of the q's alone. For each q in $\mathfrak{M}$ let T_q^0

denote the contravariant second-order tensor which takes w and w′ from V_q^* into $\widetilde{T}_q^{-1}(w)(w')$. Then we see that H has the following form:

$$H(q,w) = \frac{1}{2} T_q^0(w,w) + \mathcal{V}(q)$$

where $\mathcal{V}$ is some C_∞ function on $\mathfrak{M}$. Of course $\mathcal{V}$ is uniquely determined by H as $H(q,0)$. Hence H determines T^0 and hence T. Thus specifying T and $\mathcal{V}$ completely determines the motion of the system.

Assumptions I, II, and III are easily seen, in their context, to be equivalent to assuming that T not only exists but has the special form

$$\sum_{j=1}^{3n} M_j v_j v_j'$$

where the M_j are constants. Just as the M_j could not be shown to exist uniquely, so several different T's may work for the same system. In other words, two quite different Hamiltonian functions H may lead to exactly the same one-parameter group in $\mathfrak{M}_V$. On the other hand, there are many important cases in which T is unique up to a multiplicative constant. Moreover, whenever there is more non-uniqueness than this it can be removed by letting the system interact with other systems. Essential non-uniqueness of T only occurs when one has isolated too small a portion of the physical world.

Our fundamental law can be formulated in an equivalent manner without introducing the manifold $\mathfrak{M}_{V^*}$. Let $\mathcal{L}$ be any C_∞ function on $\mathfrak{M}_V$. Let $q: t \to q(t)$ be a C_∞ curve in $\mathfrak{M}$. Then if $a \leq b$ are in the domain of q we may form $\int_a^b \mathcal{L}(q(t), q'(t))\, dt$. If q gives a stationary value to this integral (in the usual sense of the calculus of variations) with respect to all competing q's with the same end points we shall say that q is an *extremal* of $\mathcal{L}$. It is a theorem, whose proof we shall omit, that the projections in $\mathfrak{M}$ of the orbits of our Hamiltonian dynamical group U are extremals of the function $\mathcal{L}$ defined as follows:

$$\mathcal{L}(q,v) = \frac{1}{2} T_q(v,v) - \mathcal{V}(q)$$

Conversely, let U_t be any C_∞ one-parameter group of automorphisms of $\mathfrak{M}_V$ and let $\mathcal{L}$ be a C_∞ function on $\mathfrak{M}_V$ of the form

$$\mathcal{L}(q,v) = \frac{1}{2} T_q(v,v) - \mathcal{V}(q)$$

where T is a Riemannian metric on $\mathfrak{M}$ and $\mathcal{V}$ a real-valued function on $\mathfrak{M}$. If the projection on $\mathfrak{M}$ of the orbits of U are extremals of $\mathcal{L}$, then $t \to U_t^T$ has a globally Hamiltonian infinitesimal generator. Indeed we may take $\frac{1}{2} T_q^0(w,w) + \mathcal{V}(q)$ as the Hamiltonian function on $\mathfrak{M}_{V^*}$. $\mathcal{L}$ is called the Lagrangian of the system. This formulation of the fundamental restriction on U in terms of the stationary properties of $\mathcal{L}$ is, of course, a more general form of Hamilton's principle already mentioned in Sec. 1-2.

Of course we may consider the stationary properties of integrals of the form $\int_a^b \mathcal{L}(q(t), q'(t))\, dt$, where $\mathcal{L}$ is a general C_∞ function on $\mathfrak{M}_V$ and not necessarily quadratic on the V_q. Also, one certainly has globally Hamiltonian C_∞ vector fields on $\mathfrak{M}_{V^*}$, where the corresponding H is not quadratic on the V_q^*. It turns out that there is a correspondence between extremals of calculus of variations problems and integral curves of globally Hamiltonian C_∞ vector fields which is considerably more general than that indicated above. We shall content ourselves here with describing the passage from $\mathcal{L}$ to H in the special case in which there is a corresponding one-to-one map of $\mathfrak{M}_V$ on $\mathfrak{M}_{V^*}$. Consider first of all a finite dimensional vector space V. As a C_∞ manifold it is its own tangent space at every point. Thus any C_∞ covariant vector field on V is at the same time a C_∞ mapping from V to V^*. Let $\mathcal{L}$ be any C_∞ real-valued function on V such that the C_∞ covariant vector field $d\mathcal{L}$ is one-to-one from V onto V^*. It is not difficult to show that $(d\mathcal{L})^{-1}$ is of the form $d\mathcal{L}^0$ for some C_∞ function $\mathcal{L}^0$ on V^* and that the arbitrary constant in $\mathcal{L}^0$ can be chosen so that $\mathcal{L}^0(\ell) = \ell(\psi(\ell)) - \mathcal{L}(\psi(\ell))$, where $\psi = (d\mathcal{L})^{-1}$. Of course we identify V with V^{**}. Moreover $\mathcal{L}^{00} = \mathcal{L}$. Thus $\mathcal{L} \to \mathcal{L}^0$ sets up a one-to-one correspondence between certain C_∞ functions on V and certain C_∞ functions on V^*. Now let $\mathcal{L}$ be a C_∞ function on $\mathfrak{M}_V$ and suppose that for each fixed q in $\mathfrak{M}$, $\mathcal{L}_q$ as a function on V_q satisfies the condition described above. Then we may form $(\mathcal{L}_q)^0$ and get a function on V_q^*. $q,w \to (\mathcal{L}_q)^0(w)$ is then a C_∞ function on $\mathfrak{M}_{V^*}$ which we denote by $\mathcal{L}^0$. As before $\mathcal{L}^{00} = \mathcal{L}$. It is not hard to see that when $\mathcal{L}$ is the Lagrangian of a dynamical system then $\mathcal{L}^0$ is the Hamiltonian as described above.

When our basic Riemannian metric T has been used to map $\mathfrak{M}_V$ onto $\mathfrak{M}_{V^*}$ so that $\mathfrak{M}_{V^*}$ may be regarded as the space of all states, we shall refer to $\mathfrak{M}_{V^*}$ as the *phase space* of the system. Real-valued functions on $\mathfrak{M}_{V^*}$ will be called *observables* or *dynamical variables*. It is now easy to generalize the remaining considerations of Sec. 1-2. $\mathcal{V}(q)$, $\frac{1}{2} T_q(w,w)$, and $\mathcal{V}(q) + \frac{1}{2} T_q(w,w)$ are observables which are called, respectively, the potential energy, the kinetic energy, and the total energy of the system. An observable that is constant on the orbits of the dynamical group U (now regarded as acting on $\mathfrak{M}_{V^*}$) is an integral. A C_∞ observable f is an integral if and only if $[f,H] = 0$. In particular $[H,H] = 0$, so that the total energy $H = \mathcal{V} + \frac{1}{2}T$ is an integral. More generally, if $\widetilde{d\phi}$ is the infinitesimal generator of a one-parameter group U^ϕ of automophisms of $\mathfrak{M}_{V^*}$, then ϕ is an integral if and only if H is a constant on the orbits of U^ϕ.

Let $t \to A_t$ be any C_∞ one-parameter group of automorphisms of $\mathfrak{M}$ and let L^A denote its infinitesimal generator. Each A_t is an automorphism of $\mathfrak{M}$, so $(dA_t)_q$ maps V_q linearly on $V_{A_t(q)}$ and $(dA_t)_q^{*-1}$ maps V_q^* linearly on $V^*_{A_t(q)}$. Thus $q,w \to A_t(q), (dA_t)_q^{*-1}(w)$ is a one-to-one map $\overset{0}{A}_t$ of $\mathfrak{M}_{V^*}$ on $\mathfrak{M}_{V^*}$. Clearly $t \to \overset{0}{A}_t$ is a one-parameter group of automorphisms of $\mathfrak{M}_{V^*}$. We call it the group generated by A_t. Consider L^A. It assigns a vector v to each q in $\mathfrak{M}$. This v may be regarded as a function (which is linear) on V_q^*. Thus L^A defines a function on $\mathfrak{M}_{V^*}$, $q,w \to w(L_q^A)$. Call this function ϕ^{L^A}. It is a theorem, whose proof we leave to the reader, that $\widetilde{d\phi^{L^A}} = L^{\overset{0}{A}}$. In other words, the infinitesimal generator of the group $\overset{0}{A}$ generated by A is globally Hamiltonian, and the function on $\mathfrak{M}_{V^*}$, whose differential defines it, is ϕ^{L^A}. Functions in $\mathfrak{M}_{V^*}$ linear on each V_q^* are called generalized momenta. We get an integral linear on the V_q^* from each one-parameter group A in $\mathfrak{M}$ that leaves T and $\mathcal{V}$ fixed. These integrals as before are called momentum integrals.

It follows from the preceding that to describe an actual physical system it suffices to describe the space $\mathfrak{M}$ of all configurations and to give a Riemannian metric T on $\mathfrak{M}$ and a potential function $\mathcal{V}$. The infinitesimal generator of the dynamical group can be written down at once and predicting the future of the system from the present is reduced to the (often very difficult) problem of integrating a system of ordinary differential equations. We conclude this section with a few examples.

In celestial mechanics $\mathfrak{M}$ is the Euclidean 3n-dimensional space of all coordinates of the n planets under consideration.

$$T(x_1,y_1,z_1,\ldots,x_n,y_n,z_n, v_1,\ldots,v_{3n}) = m_1(v_1^2 + v_2^2 + v_3^2) + \cdots + m_n(v_{3n-2}^2 + v_{3n-1}^2 + v_{3n}^2)$$

$$\mathcal{V}(x_1,y_1,z_1,\ldots,x_n,y_n,z_n) = \cdot\cdot - \sum_{i,j=1}^{n}{}_{i \neq j} G \frac{m_i m_j}{\sqrt{(x_i - x_j)^2 + (y_i - y_j)^2 + (z_i - z_j)^2}}$$

where the m_i and G are constants determined from experiment. Strictly speaking the theory formulated above does not apply to this example, as $\mathcal{V}$ has singularities. However, it can be made to apply by modifying $\mathcal{V}$ for very small values of the distances so as to make collision impossible. When $n = 2$ the resulting differential equations can be integrated completely, and one finds that each planet moves in a fixed orbit about its common center of gravity. This orbit is always an ellipse, a parabola, or a hyperbola—which one it is depending on the initial conditions. For higher n, only rather special partial results are known.

In the everyday mechanics of inclined planes, levers, pulleys, tops, etc., one has again a finite system of mass particles with coordinates $x_1, y_1, z_1, \ldots, x_n, y_n, z_n$ but now there are certain constraints, so that $\mathfrak{M}$ is a C_∞ submanifold of E^{3n}. If the z axis is chosen perpendicular to the surface of the earth, then $\mathcal{V}$ is the restriction to $\mathfrak{M}$ of $m_1 z_1 + \cdots + m_n z_n$ and T the restriction to $\mathfrak{M}_V^*$ of $m_1(v_1^2 + v_2^2 + v_3^2) + \cdots + m_n(v_{3n-2}^2 + v_{3n-1}^2 + v_{3n}^2)$. In the theory of a simple "rigid body" $\mathfrak{M}$ is a 6-dimensional manifold obtained by taking the intersection of $n(n-1)/2$ $3n-1$ dimensional hypersurfaces of the form

$$(x_i - x_j)^2 + (y_i - y_j)^2 + (z_i - z_j)^2 = d_{ij}$$

where each d_{ij} is a constant.

1–4 Oscillations, Waves, and Hilbert Space

Let q,w be a point on the phase space of a dynamical system which is stationary, i.e., such that $U_t(q,w) = q,w$ for all t. This will be the case if and only if $w = 0$ and $(d\mathcal{V})_q = 0$. Points of $\mathfrak{M}$ at which $d\mathcal{V}$ is zero are called *equilibrium points* of the system. Now the condition $d\mathcal{V}_q = 0$ is necessary (but not sufficient) in order that $\mathcal{V}$ have either a local maximum or a local minimum at q. To decide which one it has (if either) one looks at the bilinear form B_q in V_q whose matrix in a coordinate system is $\partial^2\mathcal{V}/\partial q_i \partial q_j$. [To define it invariantly form $L(M(\mathcal{V}))$. The value at q depends only on L_q and M_q.] If $B_q(v,v) > 0$ except when $v = 0$, the form is said to be positive definite and q is a local minimum for $\mathcal{V}$. In that case the kinetic energy will be larger at q than at any nearby point, and if the system is slightly disturbed from its equilibrium position at q it will not be able to go far without having its kinetic energy, and hence its velocity vector, drop to zero. Accordingly q will be said to be a point of *stable* equilibrium. If motion starts near such a point it will stay near it. Now near any point a manifold may be approximated by its tangent space. Moreover if $\mathfrak{M}$ is a vector space with q at 0 we may approximate T near 0 by a tensor with constant coefficients and $\mathcal{V}$ by the quadratic tensor in its Taylor formula. In other words, the "small" oscillations of an arbitrary system near a position of stable equilibrium may be studied by studying the motions of a system having the following properties:

(1) $\mathfrak{M}$ is a finite-dimensional vector space
(2) $\mathcal{V}$ is a positive definite quadratic form
(3) T_q is independent of q. Such a mechanical system we shall call a *linear system*.

Let $\mathfrak{M}$, $\mathcal{V}$, T be any linear system. Then V_q at each point may be identified with $\mathfrak{M}$ in a natural way so that $\mathfrak{M}_V = \mathfrak{M} \oplus \mathfrak{M}$ and is also a vector space. Similarly $\mathfrak{M}_{V^*} = \mathfrak{M} \oplus \mathfrak{M}^*$. Since $\mathfrak{M}_{V^*}$ is a vector space it also may be identified with its tangent space at every point, and the covariant tensor field dW^0 will be an assignment to each point of $\mathfrak{M}_{V^*}$ of an anti-symmetric bilinear functional on $\mathfrak{M}_{V^*}$. We compute easily that dW^0 is constant and takes the vectors x_1, ℓ_1, x_2, ℓ_2 into $\ell_1(x_2) - \ell_2(x_1)$. Equipped with the positive definite bilinear form T as an inner product, $\mathfrak{M}$ becomes a finite dimensional, real Hilbert space. Let $\mathcal{V}^0(\phi, \psi)$ denote the bilinear form such that $\mathcal{V}^0(\phi, \phi)$ is the potential energy V. Then for each fixed ϕ in $\mathfrak{M}, \psi \to \mathcal{V}^0(\psi, \phi)$ is linear in ψ. Hence there exists a vector ϕ' such that $T(\psi, \phi') \equiv \mathcal{V}^0(\phi, \psi)$. Clearly ϕ' depends linearly on ϕ, so that we may set $\phi' = F(\phi)/2$, where F is a uniquely determined linear transformation of $\mathfrak{M}$ into $\mathfrak{M}$. Since $\mathcal{V}(\phi) = T(F(\phi), \phi)/2 = T(\phi, F(\phi))/2$, we see that our system is determined by giving T and the transformation F. Since $T(F(\phi), \psi) = T(F(\psi), \phi) = T(\phi, F(\psi))$ we see that F is *self-adjoint* with respect to the inner product T. By a standard theorem in the theory of linear algebras there exists a basis $\phi_1, \ldots, \phi_n$ for $\mathfrak{M}$ such that $T(\phi_i, \phi_j) = \delta_i^j$ and $F(\phi_j) = \lambda_j \phi_j$, where the λ_j are real and positive. The positivity, of course, follows from the positive definiteness of $T(F(\phi), \psi)$. In the coordinate system for $\mathfrak{M}$ defined by this basis we have

$$\mathcal{V}(q_1, \ldots, q_n) = \frac{1}{2} \lambda_1 q_1^2 + \cdots + \lambda_n q_n^2$$

and

$$H(q_1, \ldots, q_n, p_1, \ldots, p_n) = p_1^2/2 + \cdots + p_n^2/2 + \frac{1}{2}(\lambda_1 q_1^2 + \cdots + \lambda_n q_n^2)$$

$$= (p_1^2 + \lambda_1 q_1^2)/2 + \cdots + (p_n^2 + \lambda_n q_n^2)/2$$

Thus Hamilton's equations are

$$\dot{p}_i = -\lambda_i q_i$$

$$\dot{q}_i = p_i$$

Thus $\ddot{q}_i = -\lambda_i q_i$ and $q_i = a_i \sin(\sqrt{\lambda_i} t - \theta_i)$, where the a_i and θ_i depend upon the initial conditions. We draw several conclusions:

1. The coordinates q_i vary independently of one another: The system acts as an assembly of n independent systems each with a one-dimensional configuration space.

2. The coordinate q_i oscillates from $-a_i$ to a_i and back again with a frequency ν_i independent of the initial conditions, and equal to $\sqrt{\lambda_i}/2\pi$.

3. Changing the initial conditions can change only the amplitudes a_i of the oscillation and the time (the phase) at which it passes through its maximum and minimum. It does not change the "simple harmonic" form of the oscillation. The frequencies $\nu_i = \sqrt{\lambda_i}/2\pi$ are called the "fundamental" or "characteristic" frequencies of the system.

It is to be emphasized that the "normal coordinates" q_i that execute independent simple harmonic motions may be related in no simple way to the rectangular coordinates of the original particles. On the other hand, the total energy of the original system in the state whose normal coordinates are $q_1, \ldots, q_n, p_1, \ldots, p_n$ is equal to

$$\frac{1}{2} \sum_{j=1}^{n} (p_j^2 + \lambda_j q_j^2) = \frac{1}{2} \sum_{j=1}^{n} \lambda_j a_j^2 [\sin^2 (\sqrt{\lambda_j} t - \theta_j) + \cos^2 (\sqrt{\lambda_j} t - \theta_j)]$$

$$= \sum_{j=1}^{n} 2\pi^2 \nu_j^2 a_j^2$$

and depends only on the *amplitudes* of motion of the normal coordinates. We note in addition that on each orbit of the motion not only is the total energy constant but also the part of it, $2\pi^2\nu_j^2 a_j^2$, associated with each particular normal coordinate. We may thus speak of the "spectral distribution" of the energy of the system on each particular orbit (or state of motion). This is often expressed as a *measure* on the positive real axis—the measure assigned to the subset E being

$$\sum_{\nu_j \in E} 2\pi^2 \nu_j^2 a_j^2$$

Now let us return to our linear system as given and discuss its motion globally without recourse to normal coordinates. From the equations $\dot{q}_i = p_i$, $\dot{p}_i = -\lambda_i q_i$, we see at once that $\ddot{q}_i = -\lambda_i q_i$, which may be formulated in coordinate-free form as $\ddot{\phi} = -F(\phi)$. This equation could also, of course, be derived from any coordinate system or even *a priori*. The corresponding first-order equation in $\mathfrak{M}_V = \mathfrak{M} \oplus \mathfrak{M}$ is $d/dt\,[\phi, \theta] = (\theta, -F(\phi))$. When a vector space is regarded as a C_∞ manifold its contravariant vector fields are simply functions mapping the vector space into itself. Thus it makes sense to say that a contravariant vector field is linear. From the last equation we see that the dynamical group in $\mathfrak{M} \oplus \mathfrak{M}$ has the *linear* infinitesimal generator $(\phi, \theta) \to \theta, -F(\phi)$. It follows at once (since the canonical map of $\mathfrak{M}_V$ into $\mathfrak{M}_{V^*}$ is linear) that the dynamical group in $\mathfrak{M}_{V^*} = \mathfrak{M} \oplus \mathfrak{M}^*$ is also linear.

Let X be any finite dimensional, real vector space and let A be any linear transformation of X into Y. For any polynomial P with real coefficients we may form P(A) by formal substitution and verify that $P \to P(A)$ is a homomorphism of the ring of all polynomials with a subring of the ring of all linear transformations of Y into X. More generally, if f is a real function with an everywhere convergent power series expansion $a_0 + a_1x + a_2x^2 + \cdots$ it is easy to see that the series $a_0 + a_1A + a_2A^2 + \cdots$ also converges and gives us a well-defined operator f(A). $f \to f(A)$ is again a ring homomorphism. In particular e^{tA} makes sense for all real t and satisfies the identity $e^{(t_1+t_2)A} = e^{t_1A}e^{t_2A}$. Thus $t \to e^{tA}$ is a one-parameter group of non-singular linear transformations of X into X. It is also a C_∞ group of automorphisms of X regarded as a C_∞ manifold, and it is easy to see that its infinitesimal generator is the *linear* contravariant vector field $\phi \to A(\phi)$. Thus in $t \to e^{At}$ we have an explicit integration of the differential equation defined by any linear contravariant vector field. It is almost obvious that any C_∞ one-parameter group of non-singular linear transformations has a linear infinitesimal generator, and it is true, in general, that a one-parameter group of linear transformations which is as much as measurable in t is actually C_∞. In the case at hand our dynamical group preserves a certain positive definite inner product: $H(\phi, \theta) = \frac{1}{2}T(F(\phi), \phi) + \frac{1}{2}T(\theta, \theta)$. In general we may ask for the conditions under which e^{At} preserves an inner product (ϕ, ψ) in X. As is shown in the elementary theory of vector spaces, $(B\phi, B\psi) = (\phi, \psi)$ for all ϕ and ψ if and only if $B^* = B^{-1}$, and a transformation B having this property is said to be orthogonal (with respect to the given inner product). Now in the operational calculus $f(A^*) = f(A)^*$, so e^{At} is orthogonal if and only if $e^{A^*t} = (e^{At})^{-1} = e^{(-A)t}$. Thus $t \to e^{At}$ is a one-parameter group of orthogonal transformations if and only if $A^* = -A$, that is, A is *skew-adjoint*. Let A be any non-singular skew-adjoint operator; then A^2 is self-adjoint. Moreover $(-A^2\phi, \phi) = (-A\phi, A^*\phi) = (A^*\phi, A^*\phi) > 0$. Thus $-A^2$ is not only self-adjoint but has only positive characteristic values. Thus $-A^2 = B^2$, where B is self-adjoint, has only positive characteristic values, and commutes with everything that commutes with A. Let $J = AB^{-1}$; then $J^2 = -I$, $J^* = -J$, and J commutes with A. Let us make X into a complex vector space by letting $i\phi = J(\phi)$. Then A is complex linear. Let $[\phi, \psi] = (\phi, \psi) - i(J\phi, \psi)$. Then

$$[\phi, \psi] = \overline{[\psi, \phi]}, \quad [\phi, \phi] = (\phi, \phi) > 0 \qquad \text{for } \phi \neq 0$$

and

$$[i\phi, \psi] = (J\phi, \psi) - i(J^2\phi, \psi) = i(\phi, \psi) + (J\phi, \psi)$$

$$= i((\phi, \psi) - i(J(\phi), \psi)$$

Thus $[\phi, \psi]$ is an inner product which makes X into a *complex*, finite dimensional Hilbert space. Since $(Je^{At}\phi, e^{At}\psi) = (e^{At}J\phi, e^{At}\psi) = (J\phi, \psi)$, e^{At} leaves the form $[\phi, \psi]$ invariant and is therefore a one-parameter group of *unitary transformations*. The infinitesimal generator A is skew-adjoint with respect to [ ] and now we may write $A = iB$, where B is self-adjoint. To summarize: The infinitesimal generator of a one-parameter group of orthogonal transformations in a finite dimensional, real Hilbert space is a skew-adjoint linear transformation A. If A is non-singular we can convert X into a complex Hilbert space in such a manner that the old real inner product is the real part of the new complex inner product and A is still skew-adjoint and linear. The multiplication by i can be chosen so that $-iA$ is a positive definite self-adjoint operator, and if this is done the complex structure is uniquely determined.

We may regard our linear dynamical system as determined by the giving of a real, finite dimensional Hilbert space $\mathfrak{M}$ and a self-adjoint operator F—the Hamiltonian then being the function $\phi, \theta \to \frac{1}{2}(\theta,\theta) + \frac{1}{2}(F(\phi), \phi)$ in $\mathfrak{M} \oplus \mathfrak{M}$. On a real Hilbert space we may, of course, forget the difference between $\mathfrak{M}$ and $\mathfrak{M}^*$. The skew-adjoint infinitesimal generator A of the dynamical group then takes (ϕ,θ) into $(\theta, -F(\phi))$, so its square takes ϕ, θ into $-F(\phi)$, $-F(\theta)$. Thus $-A^2$ is just F extended from $\mathfrak{M}$ to $\mathfrak{M} \oplus \mathfrak{M}$, in the obvious fashion. If $\phi_1, \ldots, \phi_n$ are mutually orthogonal members of $\mathfrak{M}$ such that $F(\phi_j) = \lambda_j \phi_j$, then $B^2(\phi_j, 0) = \lambda_j(\phi_j, 0)$ and $B(\phi_j, 0) = \sqrt{\lambda_j}(\phi_j, 0)$. Moreover the $\phi_j, 0$ are $\perp$ in $\mathfrak{M} \oplus \mathfrak{M}$ regarded as a complex Hilbert space. Thus the self-adjoint operator $-iA$ has just the numbers $\sqrt{\lambda_j} = 2\pi\nu_j$ as eigenvalues. It follows that we may convert the phase space of our system into a complex Hilbert space in such a manner that

(a) The value of the Hamiltonian at an element is one-half the square of the norm.

(b) The dynamical group is a one-parameter unitary group.

(c) When this group is put in the form $t \to e^{2\pi itB}$, the self-adjoint operator B has the fundamental frequencies of the system as eigenvalues.

When we attempt to extend the considerations of C to systems with an infinite number of degrees of freedom—vibrating strings, membranes, etc., one meets difficulties of a technical character arising from the lack of a suitably general uniqueness theorem for partial differential equations, as well as from the lack of obvious "natural" topologies in the infinite dimensional spaces that occur.

On the other hand, the relatively highly developed theory of linear operators in infinite dimensional vector spaces makes possible a rather complete generalization of the results obtained so far in this section. The resulting theory applies to the "small" vibrations of continua as well as to the electromagnetic field and is formally extremely closely related to quantum mechanics itself.

We begin by outlining without proof some of the principal features in the mathematical theory of linear operators in Hilbert space. Let X be a real or complex vector space (no longer necessarily finite dimensional) and let (ϕ, ψ) be an inner product in X, that is, a function of two variables linear in the first such that $(\phi, \phi) > 0$ for $\phi \neq 0$ and $(\phi, \psi) = \overline{(\psi, \phi)}$. We shall suppose that X is a *separable Hilbert space* in the sense that the metric $\rho(\phi, \psi) = \sqrt{(\phi - \psi, \phi - \psi)}$ makes it into a complete and separable metric space. These conditions are, of course, automatically satisfied when X is finite dimensional. It will be important to consider linear operators that are not everywhere defined and are not continuous. However we shall usually only consider those whose domains have all X as their closure (i.e., are dense). It is trivial to verify that a linear operator is continuous if and only if it is bounded in the sense that $\|T(\phi)\|/\|\phi\|$ is bounded as ϕ varies over the non-zero elements. We thus speak interchangeably of bounded linear operators and continuous linear operators. When a bounded linear operator has a dense domain then it has a unique bounded extension to all X. Unless the contrary is stated explicitly it will be assumed that all bounded linear operators are everywhere defined.

For fixed ψ in X, $\phi \to (\phi, \psi)$ is a continuous linear function from X to the complex numbers—a so-called continuous linear functional. It may be proved that, conversely, every continuous linear functional can be so defined by a unique member of X. Thus, if we define $\overline{X}$ to be the set of all *continuous* members of the algebraic dual X^* of X, then we have a natural, one-to-one map of X on $\overline{X}$. Let T be any bounded linear operator in X; then $(T(\phi), \psi)$ is continuous and linear as a function of ϕ for each ψ in X. Hence for each ψ there exists a unique ψ^* such that $(T(\phi), \psi) = (\phi, \psi^*)$ for all ϕ. ψ^* depends linearly upon ψ, and this linear transformation is called T^*. It is easy to see that T^* is bounded and that $T^{**} = T$. As in the finite dimensional case, T is said to be *self-adjoint* if $T^* = T$ and *skew-adjoint* if $T^* = -T$. Again, as in the finite dimensional case, T is one-to-one onto and norm preserving if and only if $TT^* = T^*T = I$, i.e., if and only if $T^* = T^{-1}$. Such a T is said to be *unitary*.

It turns out to be important to consider a generalization of the notion of self-adjointness (and skew adjointness) in which the operator is not required to be bounded nor to be everywhere defined. Let T be linear and defined on a dense subspace D of X. Then $(T(\phi), \psi)$ will be defined for all ψ in X and all ϕ in D. It will not *necessarily* be continuous in ϕ for fixed ψ because T is not continuous. On the other hand, it *may* be continuous for some ψ even though T is not continuous. Let D^* denote the set of all ψ for which $(T(\phi), \psi)$ is continuous in ϕ. For each $\psi \in D^*$ this continuous linear functional defined in D will have a unique continuous extension to X and hence will be defined by a member ψ^* of X; that is, $(T(\phi), \psi) = (\phi, \psi^*)$. As before, we denote ψ^* by $T^*(\psi)$ but note that T^* need not be defined everywhere. If $D^* = D$ and $T(\phi) = T^*(\phi)$ for all ϕ in $D = D^*$ we say that T is *self-adjoint*. (The term hypermaximal is also sometimes used.) Skew-adjointness is defined analogously. Note that the condition $(T(\phi), \psi) = (\phi, T(\psi))$ for all ϕ and ψ in D does not imply self-adjointness. It implies only that $D \subseteq D^*$ and $T(\phi) = T^*(\phi)$ for all ϕ in D. Such a T is said to be *symmetric*. It may or may not be possible to enlarge the domain of a symmetric operator so as to make it self-adjoint and if it is possible it may be possible in several ways. A symmetric operator with a unique self-adjoint extension is said to be *essentially self-adjoint*. If X is $\mathcal{L}^2(-\infty, \infty)$ and T is the operator that carries each continuously differentiable function which vanishes outside of a finite interval into i times its derivative, then T is symmetric but not self-adjoint. T becomes self-adjoint by extending to include all absolutely continuous functions whose derivatives lie in $\mathcal{L}^2(-\infty, \infty)$. It can be proved that a self-adjoint operator whose domain is the whole of X is necessarily bounded.

By the orthogonal complement $M^\perp$ of a closed subspace M of X we mean the set of all θ such that $(\theta, \phi) = 0$ for all $\phi \in M$. It can be shown that $M^{\perp\perp} = M$ and that every θ in X can be written uniquely in the form $\theta_1 + \theta_2$, where $\theta_1 \in M$ and $\theta_2 \in M^\perp$. It is clear that θ_1, the "component" of θ in M, is a linear function of θ. We write $\theta_1 = P_M(\theta)$. As the reader may prove without difficulty, P_M is a bounded self-adjoint operator which is "idempotent" in the sense that $P_M^2 = P_M$. Conversely, let P be any bounded self-adjoint idempotent transformation. Then the range of P is identical with the null space of $I - P$ and is a closed subspace M of X. Since $\phi = P(\phi) + (\phi - P(\phi))$ and $((\phi - P(\phi)), P(\theta)) = ((P(\phi) - P^2(\phi)), \theta) = 0$, we see that $P = P_M$. Thus there is a natural one-to-one correspondence between bounded self-adjoint idempotents, on the one hand, and closed subspaces, on the other. For obvious reasons the bounded self-adjoint idempotent operator associated with a given closed subspace is called the *projection on the subspace* and a bounded self-adjoint idempotent operator is called a *projection*.

Let P_1 and P_2 be projections on M_1 and M_2, respectively. Then $M_1 \perp M_2$ (i.e., $M_1 \subseteq M_2^\perp$ and $M_2 \subseteq M_1^\perp$) if and only if $P_1 P_2 = P_2 P_1 = 0$. We say that P_1 and P_2 are orthogonal. More generally, if $P_1P_2 = P_2P_1$ then $P_1P_2 = P_3$ is itself a projection and so are $P_1 - P_3$ and $P_2 - P_3$. Moreover the projections $P_1 - P_3$, $P_2 - P_3$, and P_3 are orthogonal in pairs. More generally, if $P_1, \ldots, P_n$ are projections that commute with one another and we form products of the P_j and the $I - P_k$, we get a family $P'_1, \ldots, P'_r$ of projections that are orthogonal in pairs and are such that each P_j is uniquely a sum of the form $P'_{i_1} + \cdots + P'_{i_s}$. It follows easily that any finite family of mutually commuting projections generates a Boolean algebra of projections, where $P_1 \cap P_2 = P_1P_2$, $P_1 \cup P_2 = P_1 + P_2 - P_1P_2$, and $P' = I - P$. Let $P_1, \ldots, P_n$ now be mutually orthogonal projections such that $P_1 + \cdots + P_n = I$. Then every ϕ in X may be written uniquely as $\phi_1 + \cdots + \phi_n$, where ϕ_j is in the subspace M_j on which P_j projects. Indeed $\phi_j = P_j(\phi)$. Let $\lambda_1, \ldots, \lambda_n$ be arbitrary, distinct, real numbers. Then $A = \lambda_1 P_1 + \cdots + \lambda_n P_n$ is a bounded self-adjoint operator. It is easy to see that there exists a non-zero θ such that $A(\theta) = \lambda\theta$ for some complex number λ if and only if $\lambda = \lambda_j$ for some λ_j, and then the possible θ's are the members of the range of P_j. Thus when A can be written in the indicated form, the P_j and λ_j are uniquely determined. When X is finite dimensional every A can be so written, but this is far from being true in the infinite dimensional case even if we allow infinite sums. Let $X = \mathcal{L}^2(S, \mu)$, where μ is a measure on the space S. Let g be a bounded real-valued measurable function and let $A_g(f) = fg$. Then A_g is a bounded self-adjoint operator. Then $A_g(f) = \lambda f$ if and only if $(g - \lambda)f = 0$ almost everywhere in S. Thus, if g takes on each value only finitely many times and every point has measure zero, then $f = 0$ almost everywhere and is the zero element of X. It follows at once that there are many examples of bounded self-adjoint operators with no eigenelements whatever. On the other hand, it is easy to generalize the decomposition $A = \lambda_1 P_1 + \cdots + \lambda_n P_n$ to arbitrary self-adjoint operators A provided that we reformulate it somewhat. For each Borel† subset E of the real line let $P_E = P_{i_1} + \cdots + P_{i_r}$, where $\lambda_{i_1}, \ldots, \lambda_{i_r} \in E$ and the other λ_j are not in E. Then the assignment $E \to P_E$ completely describes the P_j and the λ_j. In fact, if $\{\lambda\}$ denotes the set whose only element is λ, then $P_{\{\lambda\}} \neq 0$ if and only if $\lambda = \lambda_j$ for some j and $P_{\{\lambda_j\}} = P_j$. Moreover $E \to P_E$ has the following properties:

(a) $P_E P_F = P_{E \cap F}$ for all E and F.

(b) $P_\phi = 0$, $P_R = I$, where R is the whole real line.

---

†The Borel sets are the members of the smallest family of sets which includes the open sets and has the following two properties:

1. The complement of any member of the family is in the family.
2. The intersection $A_1 \cap A_2 \cap \cdots$ of the members of any countable subfamily is in the family.

(c) $P_E = P_{E_1} + P_{E_2} + \cdots$ whenever $E = E_1 \cup E_2 \cup \cdots$ and $E_i \cap E_j = \phi$ for $i \neq j$.

Here $P_{E_1} + P_{E_2} + \cdots$ means the unique projection P such that $P_{E_1}\phi + P_{E_2}\phi + \cdots = P\phi$ for all ϕ or equivalently the projection on the closed linear subspace generated by the ranges of P_{E_j}. Whenever we have a function from real Borel sets to projections which has properties (a), (b), and (c) we shall say that we have a *projection-valued measure*. Let $E \to P_E$ be any projection-valued measure. Then for each vector ϕ in X, $E \to (P_E(\phi), \phi)$ will be an ordinary non-negative measure in the real line and we can form integrals $\int f(x)\, d\,\alpha(x)$, where $\alpha(E) = (P_E(\phi), \phi)$. We shall write $\int f(x)\, d(P_x(\phi), \phi)$. Now, if P comes from $\lambda_1 P_1 + \cdots + \lambda P_n$ as indicated above, then $\int x\, d(P_x(\phi), \phi)$ is easily seen to be equal to $\lambda_1 P_1(\phi) \cdot \phi + \cdots + \lambda_n P_n(\phi) \cdot \phi = A(\phi) \cdot \phi$. Thus $A(\phi) \cdot \phi = \int x\, d(P_x(\phi), \phi)$, and since A is uniquely determined by the quadratic form $\phi \to A(\phi) \cdot \phi$, we have a formula expressing A in terms of P without introducing the P_j.

Now let $E \to P_E$ be any projection-valued measure that is *bounded* in the sense that P_J is the identity for some finite interval J. We may define a quadratic form $B(\phi, \phi)$ in X by setting $B(\phi, \phi) = \int x\, d(P_x(\phi), \phi)$ for all ϕ in X, and it is not difficult to prove that there exists a unique bounded self-adjoint operator A^P in X such that $(A^P(\phi), \phi) = B(\phi, \phi) = \int x\, d(P_x(\phi), \phi)$ for all ϕ in X. More generally one can show that if P is an unbounded projection-valued measure then there exists a unique unbounded self-adjoint operator A^P whose domain is the set of all ϕ for which $\int x\, d(P_x(\phi), \phi) < \infty$ and such that for all ϕ in this domain $(A^P(\phi), \phi) = \int x\, d(P_x(\phi), \phi)$. The spectral theorem is the converse of this proposition: Every self-adjoint operator is of the form A^P for a uniquely determined projection-valued measure P. It is not hard to show that A^P is bounded if and only if P is bounded. In the special case in which X has a basis of orthogonal elements $\phi_1, \phi_2, \ldots$ each of which is an eigenelement of A, $A(\phi_j) = \lambda_j \phi_j$, then P is concentrated in the set $\{\lambda_1, \lambda_2, \ldots\}$; that is, $P_E = I$ if $E = \{\lambda_1, \lambda_2, \ldots\}$. Conversely, if there exists a countable set $E = \{\lambda_1, \lambda_2, \ldots\}$ such that $P_E = I$, then X has a basis of eigenelements of A whose eigenvalues are the λ_j. When this is the case A is said to have a *pure point spectrum*.

It is often useful to look at the spectral theorem in a somewhat different way. In the special case in which $X = \mathcal{L}^2(S, \mu)$ and $A(f) = gf$ for some real measurable g, the proof of the spectral theorem is relatively trivial. Indeed it is straightforward to verify that the associated projection-valued measure P is simply the one that assigns to the set E the projection $f \to \psi f$, where $\psi(s) = 1$ if $g(s) \in E$ and $\psi(s) = 0$ if $g(s) \notin E$. On the other hand, this case is not as special as it looks. It can be proved that given any self-adjoint A in X there exists a linear isometry V of X onto some $\mathcal{L}^2(S, \mu)$ of such a nature that VAV^{-1} is of the form $f \to fg$. This statement can be viewed as a strengthened form of the spectral theorem, inasmuch as it is easy to deduce the spectral theorem from it and not quite so easy to go in the reverse direction. The reader will find it instructive to convince himself of the truth of this form of the spectral theorem in the special case of a pure point spectrum.

Given the spectral theorem it is easy to assign a meaning to $g(A)$ whenever g is a real-valued Borel function of a real variable and A is self-adjoint. Let P be the projection-valued measure associated with A. For each Borel set on the real line let $P'_E = P_{g^{-1}(E)}$. Then $E \to P'_E$ is a projection-valued measure and we define $g(A)$ to be the corresponding self-adjoint operator. It can be shown that $g \to g(A)$ preserves sums and products and that $g(A)$ is bounded if and only if g is bounded on the complement of a set E such that $P_E = 0$. In particular $\sin tA$ and $\cos tA$ are bounded for all real t and all self-adjoint A. Thus $\cos tA + i \sin tA = e^{itA}$ is defined whenever X is a complex Hilbert space. It is easy to see that e^{itA} is a one-parameter unitary group which is continuous in the sense that $(e^{itA}\phi, \psi)$ is a continuous function of t for all ϕ and ψ in X. Conversely, let $t \to U_t$ be any continuous one-parameter unitary group. By a fundamental theorem due to M. H. Stone there exists a unique self-adjoint operator A such that $e^{itA} = U_t$ for all t. Thus, if we admit unbounded self-adjoint operators, the one-to-one correspondence between one-parameter unitary groups and self-adjoint operators is preserved. When X is a real Hilbert space, multiplication by i is meaningless so that we cannot form e^{itA}. On the other hand, if A is skew-adjoint then A^2 is self-adjoint so we can form $g(A)$ whenever g is an even function. But any g is of the form $h_1 + xh_2$, where h_1 and h_2 are even, so we may set $g(A) = h_1(A) + Ah_2(A)$. In particular it is possible to form e^{tA} and show that it is a continuous one-parameter group of orthogonal transformations. Conversely we can adapt Stone's theorem and show that every continuous one-parameter group of orthogonal transformations in a real Hilbert space is of the form e^{tA} where A is skew-adjoint. Finally, whenever A is one-to-one and skew-adjoint, one can adapt the finite dimensional argument and show that X can be made into a complex Hilbert space in such a way that A is complex linear and the old and new norms are the same.

Returning now to physics let us approach the problem of studying systems with infinite dimensional configuration spaces by examining the system obtained by viewing a countable infinity of independent systems as a single system. Let $\mathfrak{M}_1, \mathfrak{M}_2, \ldots$ be the configuration spaces of the constituent systems which for definiteness and simplicity we suppose to be linear. Then the phase space associated with $\mathfrak{M}_j$ will be $\mathfrak{M}_j \oplus \mathfrak{M}_j^*$ and the dynamical group U^j will be a one-parameter group of linear transformations $t \to U_t^j$ in $\mathfrak{M}_j \oplus \mathfrak{M}_j^*$. The configuration space of the composite system is then the set $\mathfrak{M}$ of all infinite sequences $\phi_1, \phi_2, \ldots$ with $\phi_j \in \mathfrak{M}_j$ and the motion in $\mathfrak{M}$ is defined by a one-parameter group $t \to U_t$ in the phase space Λ of all infinite sequences $\phi_1, \ell_1, \phi_2, \ell_2, \ldots$, where $\phi_j, \ell_j \in \mathfrak{M}_j \oplus \mathfrak{M}_j^*$. $U_t(\phi_1, \ell_1, \phi_2, \ell_2, \ldots)$ is, of course, just $U_t^1(\phi_1, \ell_1)$, $U_t^2(\phi_2, \ell_2), \ldots$.

So far so good. However we meet difficulties if we attempt to go beyond simply describing the motion of the system and attempt to introduce a Hamiltonian or regard Λ as the cotangent bundle of $\mathfrak{M}$. $\mathfrak{M}$ is a vector space in a natural way but $\mathfrak{M}^*$ the dual of $\mathfrak{M}$ is not the set of all sequences $\ell_1, \ell_2, \ldots$ with $\ell_j \in \mathfrak{M}_j^*$. In fact $\ell_1(\phi_1) + \ell_2(\phi_2) + \cdots$ makes sense for all $\phi_1, \phi_2, \ldots$ in $\mathfrak{M}$ if and only if $\ell_j = 0$ for all but a finite number of j. Thus Λ the natural choice for phase space, is not $\mathfrak{M} \oplus \mathfrak{M}^*$. If we had only a finite number of systems $\mathfrak{M}_j$, U^j and if H^j were the Hamiltonian function for $\mathfrak{M}_j$, U^j, then setting $H(\phi_1, \ell_1, \phi_2, \ell_2, \ldots, \phi_n \ell_n) = H^1(\phi_1, \ell_1) + H^2(\phi_2, \ell_2) + \cdots + H^n(\phi_n, \ell_n)$ would give us a Hamiltonian function for $\mathfrak{M}$, U. However in the case at hand we have an infinite number of summands and the series will not in general converge. Of course, a Hamiltonian function is determined only up to an additive constant and by suitably adjusting this constant in each H^j we may arrange matters so that the sum of the H^j converges on a subset of Λ including the trajectory of any particular point. However, there seems to be no one function H that describes the motion of the whole system. The best that we can do is to divide Λ into disjoint sets, each invariant under the motion and on each of which there is a Hamiltonian function defined.

Now to go from one subset of Λ having a Hamiltonian function to another involves changing the "total energy" of the system by an "infinite amount." Thus one could argue that only one of the disjoint subsets is physically accessible and that one should take phase space to be that subset containing $0, 0, 0, \ldots$. If this is done, configuration space becomes a subset $\tilde{\mathfrak{M}}$ of $\mathfrak{M}$ consisting of all $\phi_1, \phi_2, \ldots$ with $V'(\phi_1, \phi_1) + V^2(\phi_2, \phi_2) + \cdots < \infty$, where V^j is the inner product in $\mathfrak{M}_j$ defining the potential energy. Phase space becomes $\tilde{\mathfrak{M}} \oplus \tilde{\tilde{\mathfrak{M}}}$, where $\tilde{\tilde{\mathfrak{M}}}$ is the set of all $\ell_1, \ell_2, \ldots$ with $T^1(\ell_1, \ell_1) + T^2(\ell_2, \ell_2) + \cdots < \infty$ and T^j is the inner product in $\mathfrak{M}_j^*$ defining the kinetic energy. However $\tilde{\tilde{\mathfrak{M}}}$ still cannot be identified with the dual of $\tilde{\mathfrak{M}}$, and we cannot carry over the finite dimensional Hamiltonian formalism without making further changes.

The essential feature of the Hamiltonian (or Lagrangian) formalism of course is the fact that it provides an algorithm for passing from a single function on phase space to the infinitesimal generator of the dynamical group. We shall give such an algorithm which applies to a class of linear systems including the case at hand and to all finite dimensional linear systems. Let $\mathfrak{M}$ be a real vector space which we think of as being "dense" in our as yet undefined configuration space.

Let V and T be positive definite inner products in $\mathfrak{M}$ which we think of as defining the potential and kinetic energies, respectively. We consider the norms $\sqrt{V(\phi,\phi)}$, $\sqrt{T(\phi,\phi)}$, $\sqrt{V(\phi,\phi)} + \sqrt{T(\phi,\phi)}$ and obtain real Hilbert spaces $\mathfrak{M}_V$, $\mathfrak{M}_T$, and $\overline{\mathfrak{M}}$ by taking the completions of $\mathfrak{M}$ in these norms. Since the third norm is larger than the other two, each of these has a unique extension to $\overline{\mathfrak{M}}$. If these extensions are positive definite (and we shall consider only this case) then $\mathfrak{M}_V$ and $\mathfrak{M}_T$ may be regarded as completions of $\overline{\mathfrak{M}}$. Let $\mathcal{H}$ be the real Hilbert space $\mathfrak{M}_V \oplus \mathfrak{M}_T$ and let I be the operator defined as follows. $I(\phi)$ is defined whenever $\phi \in \overline{\mathfrak{M}} \subseteq \mathfrak{M}_T$ and $I(\phi) = \phi$, where the ϕ on the right is regarded as an element of $\mathfrak{M}_V$. Thus I is an operator (perhaps not everywhere defined), whose domain and range are dense in $\mathfrak{M}_T$ and $\mathfrak{M}_V$, respectively. Let I^* denote the adjoint of I. Then I^* is an operator from a dense subspace of $\mathfrak{M}_V$ to $\mathfrak{M}_T$ defined for all θ in $\mathfrak{M}_V$ such that $\phi \to V(I(\phi),\theta)$ is continuous in the $\mathfrak{M}_T$ norm. $I^*(\theta)$ is the unique vector θ^* such that $V(I(\phi),\theta)) = T(\phi,\theta^*)$. Note that because of the difference between T and V, I^* is not in any sense an "identity" operator. It is routine to verify that $I^{**} = I$, and that the operator A that takes ϕ,ψ in $\mathcal{H}$ into $I(\psi)$, $-I^*(\phi)$ wherever I and I^* are defined, is a skew-adjoint operator. The group $t \to e^{tA}$ acting in the phase space $\mathcal{H}$ is the dynamical group which our algorithm assigns to the "Lagrangian" V, T.

To see the connection between this algorithm and earlier considerations let us look at the unique self-adjoint operator K in $\overline{\mathfrak{M}}$ such that $(K(\phi),\phi) = V(\phi,\phi)$. Here the inner product on the left is that defining the norm in $\overline{\mathfrak{M}}$. Consider the special case in which K has a pure point spectrum. Let $\phi_1,\phi_2,\ldots$ be a basis of eigenvectors for K with $(\phi_j,\phi_j) = 1$. Then if $\phi = q_1\phi_1 + q_2\phi_2 + \cdots$ we compute that

$$V(\phi,\phi) = \sum_{j=1}^{\infty} \frac{q_j^2}{\mu_j}$$

where $K(\phi_j) = (1/\mu_j)\phi_j$, and that

$$T(\phi,\phi) = \sum_{j=1}^{\infty} \frac{q_j^2}{m_j}$$

where $(1/m_j) + (1/\mu_j) = 1$. Thus we can choose a basis for $\overline{\mathfrak{M}}$ in which both V and T are diagonal and hence consider our system as a collection of independent harmonic oscillators. It is straightforward to verify that the dynamical group we have defined using I and I^* coincides with the one we get by blending the dynamical groups of the component systems and restricting phase space to the points for which the total energy is finite.

If $\phi,\psi \in \overline{\mathfrak{M}} \oplus \overline{\mathfrak{M}}$ and U_t is the dynamical group then $U_t(\phi,\psi)$

will be differentiable at $t = 0$ and $d/dt\,(U_t(\phi,\psi))|_{t=0}\,A(\phi,\psi) = \psi, -I^*(\phi)$. In the more usual (and more elliptical) form we have $d/dt\,(\phi,\psi) = \psi, -I^*(\phi)$ or $d\phi/dt = \psi$ $d\psi/dt = -I^*(\phi)$. Thus the "configuration vector" ϕ satisfies the second-order equation $d^2\phi/dt^2 = -I^*(\phi)$. In applications, ϕ is usually a scalar- or vector-valued function on 3-space and I^* is a differential operator. When ϕ is regarded as a function of x, y, z and t instead of an $\mathfrak{M}$-valued function of t, then $d^2\phi/dt^2 = -I^*(\phi)$ becomes a partial differential equation of the form $\partial^2\phi/\partial t^2 = -I^*(\phi)$, and it is in this form that the physical law is usually stated. It is unsatisfactory in that it is incomplete. No well-defined dynamical group is described by such an equation until we have singled out a class of ϕ's and $\partial\phi/\partial t$'s for which one can prove suitable existence and uniqueness theorems.

As an example consider the mathematics of the "small" vibrations of a string that is fixed at two end points. $\mathfrak{M}$ is the space of all twice differentiable real-valued functions on $0 \leq x \leq \ell$ such that $f(0) = f(\ell) = 0$. A member of $\mathfrak{M}$ describes the displacement of this string from straightness as a function of the distance from the left-hand fixed point.

$$V(f,f) = k \int_0^{\ell} \left(\frac{df}{dx}\right)^2 dx$$

and

$$T(f,f) = \mu \int_0^{\ell} f^2\, dx$$

Here μ and k are positive constants depending upon the material of which the string is made and the "tightness" with which it has been stretched. These expressions can be derived from the laws of particle mechanics by regarding a string as a limiting case of a sequence of particles interacting in a certain way. Since

$$V(f,f) = k \int_0^{\ell} \left(\frac{df}{dx}\right)^2 dx = -k \int_0^{\ell} \left(\frac{d^2f}{dx^2}\right) f\, dx = -\frac{k}{\mu}\, T\!\left(\frac{d^2f}{dx^2},\, f\right)$$

we see that I^* restricted to $\mathfrak{M}$ is $-(k/\mu)\,(d^2f)/(dx^2)$, so the classical "equations of motion" are

$$\frac{\partial^2 f}{\partial t^2} = -\frac{k}{\mu}\frac{\partial^2 f}{\partial x^2}$$

In dealing with the vibrations of an "elastic solid" occupying a finite volume R of E^3, $\mathfrak{M}$ is the set of all twice differentiable vector-valued functions on R that on the boundary of R have values orthogonal to the boundary normal. A member $f = \ell_1, \ell_2, \ell_3$ of $\mathfrak{M}$ describes the (vector) displacement from equilibrium as a function of position in R.

$$T(f,f) = \iiint_R \mu(\ell_1^2 + \ell_2^2 + \ell_3^2)\,dx\,dy\,dz$$

$$V(f,f) = c_1 \iiint_R \left[\frac{\partial \ell_1}{\partial x} + \frac{\partial \ell_2}{\partial y} + \frac{\partial \ell_3}{\partial z}\right]^2 dx\,dy\,dz$$

$$+ c_2 \iiint_R \left[\left(\frac{\partial \ell_1}{\partial z} + \frac{\partial \ell_3}{\partial x}\right)^2 + \left(\frac{\partial \ell_2}{\partial z} + \frac{\partial \ell_3}{\partial y}\right)^2\right.$$

$$+ \left(\frac{\partial \ell_1}{\partial y} + \frac{\partial \ell_2}{\partial x}\right)^2 + 2\left(\frac{\partial \ell_1}{\partial x}\right)^2$$

$$\left. + 2\left(\frac{\partial \ell_2}{\partial y}\right)^2 + 2\left(\frac{\partial \ell_3}{\partial z}\right)^2\right] dx\,dy\,dz$$

where μ, c_1, and c_2 depend upon the material constitution of the body. We leave it to the reader to compute the classical equations of motion.

In dealing with several elastic media in contact or one in which c_1, c_2, and μ vary from point to point there is an evident generalization of the above formulas for T and V. Moreover similar formulas exist for the vibrations of liquids and gases.

Returning to the general theory we remark that just as in the finite dimensional case we may introduce a multiplication by i into phase space so that the dynamical group $t \to U_t$ becomes a one-parameter unitary group e^{itB}, where B is self-adjoint.

When B has a pure point spectrum, as it generally does for vibrations of finite media, we can solve the initial value or prediction problem for the system as follows. Let $\phi_1, \phi_2, \ldots$ be a basis of eigenvectors for B:

$$B(\phi_j) = \lambda_j \phi_j$$

Then $U_t(\phi_j) = e^{i\lambda_j t}\phi_j$. Thus if at time $t = 0$, θ is the member of $\mathfrak{M}_V \oplus \mathfrak{M}_T$ representing the state of our system, we may get the vector θ_{t_1} representing the state at a later time $t = t_1$ as follows. Expand θ in terms of the basis vectors $\phi_1, \phi_2, \ldots$,

$$\theta = c_1\phi_1 + c_2\phi_2 + \cdots$$

Then θ_{t_1} is the unique vector whose expansion coefficients are $c_1 e^{i\lambda_1 t_1}$, $c_2 e^{i\lambda_2 t_1}, \ldots$. Of course everything depends upon actually being able to make the spectral analysis of B and find the ϕ_j. In a number of classical cases this may be done quite explicitly—the ϕ_j being various well-known families of "orthogonal functions." We see in particular that when B has a pure point spectrum the motion is analyzable into independent simple harmonic motions just as in the finite case. There is the important difference, however—that there are an infinite number of such independent oscillations.

When B does not have a pure point spectrum we can apply the spectral theorem and conclude that our Hilbert space can be mapped linearly and isometrically onto some concrete Hilbert space $\mathcal{L}^2(S, \mu)$ in such a manner that B goes over into the operator $g(s) \to \lambda(s)\, g(s)$, where λ is some real-valued function on S. U_t will then go over into the operator $g(s) \to e^{i\lambda(s)t} g(s)$, and we can give the following prescription for solving the initial value problem. Given a state vector θ, find its correspondent in $\mathcal{L}^2(S, \mu)$. Call this $\hat{\theta}$. Then $\hat{\theta}(s) e^{i\lambda(s)t_1}$ will be the correspondent in $\mathcal{L}^2(S, \mu)$ of $U_{t_1}(s)$. In many cases of interest the members of $\mathfrak{M}^T \oplus \mathfrak{M}^V$ are complex-valued functions on E^3 and the operator $\theta \to \hat{\theta}$ has an inverse which is an "integral operator." That is, there exists a function K on $E^3 \times S$ such that $\theta(x, y, z) = \int K(x, y, z, s)\, \hat{\theta}(s)\, ds$ for all θ in a dense subspace of $\mathfrak{M}^T \oplus \mathfrak{M}^V$. When this happens a wide class of trajectories of our system, i.e., functions of x, y, z and t of the form $U_t(\theta)(x, y, z)$, may be written in the form: $\int K(x, y, z, s) e^{i\lambda(s)t} \hat{\theta}(s)\, ds$ and may be regarded as a "continuous superposition" of functions of the form $K(x, y, z, s_0) e^{i\lambda(s_0)t}$. When T and V are invariant under translations and rotations of space, these basic functions take the form $\exp\{i[ax + by + cz - G(a^2 + b^2 + c^2)t]\}$, where S is the set of all triples a, b, c and G depends upon the special form of B. The function $\exp\{i[ax + by + cz - G(a^2 + b^2 + c^2)t]\}$ becomes, after a suitable rotation, the function $\exp\{i[\ell x - G(\ell^2)t]\}$, where $a^2 + b^2 + c^2 = \ell^2$. For fixed x this is a simple harmonic vibration of frequency $G(\ell^2)/2\pi$. For fixed t the real and imaginary parts are cosine and sine "waves" of wavelength $2\pi/\ell$ and as t changes these "waves" move in the direction of increasing x with the velocity $G(\ell^2)/\ell$. The unrotated function is what is called a "plane wave" going in the direction of the vector a, b, c with velocity $G(a^2 + b^2 + c^2)/\sqrt{a^2 + b^2 + c^2}$ and having wavelength $2\pi/\sqrt{a^2 + b^2 + c^2}$. Thus the motion of our system may be described as a superposition of plane waves going in various directions and having various wavelengths. In general the velocity will vary with the wavelength, but there are important special cases in which $G(v) = \sqrt{v}$ and there is a unique "wave velocity." In that case the "center" of a localized disturbance in the medium in question will move with this same velocity. When the wave velocity varies with the wavelength, the center of a localized disturbance may move with a velocity different from any of the wave velocities involved in the superposition —even if these all cluster around some one value. In this case, in fact, we have $w = v - \lambda\,(dv/d\lambda)$, where $v(\lambda)$ is the wave velocity for plane waves of length λ and w is the "group" velocity. All these statements about waves can be justified by a rigorous analysis which would take us too far afield.

If we consider waves in an infinite elastic medium, the same considerations apply except that our basic plane waves are vector-valued. Analysis shows that there are two distinct cases. When the vector is perpendicular to the direction of propagation we say that we are dealing with *transverse waves* and when it is parallel to it that we are dealing with *longitudinal waves*. The wave velocity is independent of the wavelength but is different for longitudinal and transverse waves.

We conclude this section with a brief account of how its considerations may be extended to include the electromagnetic theory of light. The nature of light has been a controversial matter throughout most of the history of physics. In 1800 it seemed to be possible to explain most phenomena either on the hypothesis that it consisted of rapidly moving small particles or on the hypothesis that it was some sort of wave motion. On the other hand, neither hypothesis gave a completely satisfactory account of all phenomena. However, owing largely to the work of Young and Fresnel between 1801 and 1827, it came to be more or less generally agreed that light consisted of transverse waves in some unknown medium. In the next 40 years much work was done in an attempt to determine the nature of this medium, which was called the "ether." The fact that the waves had to be transverse was difficult to reconcile with the fact that the space through which light travels is most unlike an elastic *solid*. Moreover, according to the theory of waves in elastic solids, longitudinal waves must always be produced when transverse waves cross the boundary from one medium to another, and no experimental evidence for longitudinal light waves could be found. The latter difficulty and others were met—at least from a mathematical point of view—by MacCullagh in 1839, who showed that if we had an elastic solid whose potential energy function was of the form

$$a \iiint \left[\left(\frac{\partial \ell z}{\partial y} - \frac{\partial \ell y}{\partial z} \right)^2 + \left(\frac{\partial \ell x}{\partial z} - \frac{\partial \ell z}{\partial x} \right)^2 + \left(\frac{\partial \ell y}{\partial x} - \frac{\partial \ell x}{\partial y} \right)^2 \right] dx\, dy\, dz$$

instead of as indicated above, then waves in this solid would behave in all respects as light waves were observed to do. In particular no longitudinal waves would be produced on an interface. The physical difficulty remained and MacCullagh's work was not appreciated until much later. We mention it because of its agreement in mathematical form with the theory that was finally accepted. Note that MacCullagh's potential differs from that of an actual elastic solid in utilizing the components of the anti-symmetric part of the displacement tensor

$$\left\|\begin{matrix} \frac{\partial \ell x}{\partial x} & \frac{\partial \ell x}{\partial y} & \frac{\partial \ell x}{\partial z} \\ \frac{\partial \ell y}{\partial x} & \frac{\partial \ell y}{\partial y} & \frac{\partial \ell y}{\partial z} \\ \frac{\partial \ell z}{\partial x} & \frac{\partial \ell z}{\partial y} & \frac{\partial \ell z}{\partial z} \end{matrix}\right\|$$

instead of those of the symmetric part.

The final nineteenth-century answer to the question of the nature of light had to await the development of the theory of the electromagnetic field by Faraday and Maxwell. In 1800 one knew the facts about the attraction and repulsion of charged bodies and magnets and there were fairly highly developed (separate) mathematical theories of electrostatics and magnetostatics. However, the observation that a changing electric field generates a magnetic one and vice versa were still to come. It was only in 1800 that Volta discovered the principle of the electric battery and made steady electric currents available to experimentalists. Twenty years later Oersted discovered that a wire carrying a current could deflect a compass needle, and in 1825 Ampere published a celebrated memoir giving a complete quantitative theory of the effect. The next big step was made by Faraday, who in 1832 showed that, conversely, a current was generated in a wire moving near a magnet. In the long period of development which followed, Faraday dominated the scene until his retirement in 1855, at which time Maxwell entered the field. Both Faraday and Maxwell emphasized the "field" point of view; that is they concentrated their attention on the force fields produced by the charges, magnets, etc., instead of on these objects themselves. It turns out that the potential energy in a mechanical system in which the forces are due to charges can be expressed in the form $k/8\pi \iiint (E_x^2 + E_y^2 + E_z^2)\, dx\, dy\, dz$, where E_x, E_y, E_z are the components of the electron field and k is the so-called dielectric constant of the surounding medium. Similarly the potential energy associated with a system of magnets can be written $\mu/8\pi \iiint (H_x^2 + H_y^2 + H_z^2)$, where H_x, H_y, H_z are the components of the magnetic field and μ is the "magnetic permeability" of the medium. Maxwell and Faraday regarded an electromagnetic field as an elastic medium in a disturbed state and showed that one could regard electromagnetic phenomena as the interaction between matter and the field conceived as a mechanical system with an infinite number of degrees of freedom like an elastic solid. Maxwell made the very important theoretical advance of guessing that just as a magnetic field is produced by a current so is one produced by a changing electric field. He was led to suppose this field H to be determined by the formula curl $H = 1/c\ \partial E/\partial t$. When this is added to relationships already known one obtains the celebrated Maxwell equations governing the change in time of an electromagnetic field. In empty space away from charges and currents they take the form

$$\text{curl } E = 1/c \frac{\partial H}{\partial t} \qquad \text{curl } H = 1/c \frac{\partial E}{\partial t}$$

$$\text{div } E = 0 \qquad \text{div } H = 0$$

where c is a constant. It is easy to eliminate either E or H from these equations and we find that

$$\frac{\partial^2 H}{\partial x^2} + \frac{\partial^2 H}{\partial y^2} + \frac{\partial^2 H}{\partial z^2} = \frac{1}{c^2} \frac{\partial^2 H}{\partial t^2}$$

and

$$\frac{\partial^2 E}{\partial x^2} + \frac{\partial^2 E}{\partial y^2} + \frac{\partial^2 E}{\partial z^2} = \frac{1}{c^2} \frac{\partial^2 E}{\partial t^2}$$

From this it is easy to see that an electromagnetic disturbance is propogated in a wave-like manner with velocity c. The c turned out to be equal to the velocity of light, and Maxwell conjectured that light in fact consisted of vibrations of the electromagnetic field. This (with the modification produced by quantum mechanics) is still the accepted view.

It is easy to fit the theory of the theory of the free electromagnetic field into the framework of this section. Let $\mathfrak{M}$ denote the vector space of all twice differentiable vector-valued functions E in 3-space such that

$$1/4\pi \iiint (E_x^2 + E_y^2 + E_z^2)\ dx\ dy\ dz < \infty$$

and div $E = 0$. Let

$$V(E, E) = 1/4\pi \iiint (E_x^2 + E_y^2 + E_z^2)\ dx\ dy\ dz$$

Let

$$T(H, H) = 1/4\pi \iiint (H_x^2 + H_y^2 + H_z^2)\ dx\ dy\ dz$$

where H is the unique differentiable vector field such that curl $H = 1/c\ \partial E/\partial t$,

$$\iiint (H_x^2 + H_y^2 + H_z^2)\ dx\ dy\ dz < \infty$$

and $\operatorname{div} H = 0$. The real Hilbert space and the one-parameter orthogonal group therein defined by the procedure described on p. 40 now yield "laws of motion" for E which become Maxwell's equations when we differentiate, provided we take $\operatorname{curl}^{-1} 1/c\ \partial E/\partial t$ as the definition of H. In the "mechanical system" so defined the energy of the electric field becomes the potential energy and the energy of the magnetic field, the kinetic energy. Instead of taking E to be fundamental we may take $A = \operatorname{curl}^{-1}(E)$, where $\operatorname{div} A = 0$. Then the kinetic energy becomes

$$1/4\pi \iiint \left[\left(\frac{\partial A_y}{\partial z} - \frac{\partial A_z}{\partial y}\right)^2 + \left(\frac{\partial A_z}{\partial x} - \frac{\partial A_x}{\partial z}\right)^2 + \left(\frac{\partial A_x}{\partial y} - \frac{\partial A_y}{\partial x}\right)^2\right] dx\, dy\, dz$$

Thus we see why the formalism of MacCullagh worked.

The important point for our purposes is that our mathematical analysis of the dynamical group of a mechanical system applies equally well to the electromagnetic field since it depended only upon the infinitesimal generator of that group having a certain form. In particular, we may speak of the linear and angular momentum of the field, and when the field is confined to a finite region of space we may regard it as a system of (infinitely many) independent harmonic oscillators.

1-5 Statistical Mechanics

In dealing with the classical mechanics of a system of a very large number of particles, for example when treating a macroscopic physical system as composed of a great many very small "atoms" moving according to classical mechanical laws, it is impossible to determine the state of the system in the sense of determining all the coordinates and all the velocities of the particles. Such experiments as we can perform give us only partial information about the state and the best we can do is to give the probability that this state is in any given region of our phase space $\mathfrak{M}_V{*}$. We must broaden our concept of state to include such probability distributions and replace the mechanical problem of studying how points in $\mathfrak{M}_V{*}$ change with time by the "statistical mechanical" one of studying how probability measures in $\mathfrak{M}_V{*}$ change with time.

Let us denote by $\mathcal{O}(\mathfrak{M}_V{*})$ the set of all probability measures in $\mathfrak{M}_V{*}$, where by definition a probability measure α is a countably additive measure defined on all Borel subsets of $\mathfrak{M}_V{*}$ and such that $\alpha(\mathfrak{M}_V{*}) = 1$. If α represents the state of our system at $t = 0$ and $t \to U_t$ is the dynamical group of our system, then the probability that the system is represented by a point q in the Borel set E at the time $t = t_1$ is exactly the probability that the system was represented by a point q in the set $U_{t_1}^{-1}(E)$ at $t = 0$. Thus the state at $t = t_1$ is represented by the probability measure $E \to \alpha(U_{t_1}^{-1}(E))$. Let us define $\hat{U}_t(\alpha)$ to be the measure $E \to \alpha(U_t^{-1}(E))$. Then $t \to \hat{U}_t$ is a one-parameter group of one-to-one transformations of our generalized state space $\mathcal{O}(\mathfrak{M}_{V^*})$ onto itself. In statistical mechanics one studies $\hat{U}_t$ instead of U_t.

There is a certain natural measure in $\mathfrak{M}_{V^*}$ which we shall call the "Liouville measure" and which plays an important role in the theory. If $q_1, \ldots, q_n$ is a coordinate system for an open subset $\mathcal{O}$ of $\mathfrak{M}$ and $q_1, \ldots, q_n, p_1, \ldots, p_n$ is the corresponding coordinate system for $\pi^{-1}(\mathcal{O})$, we get a measure in $\pi^{-1}(\mathcal{O})$ by using Lebesgue measure in the image of $\pi^{-1}(\mathcal{O})$ in E^{2n} defined by this coordinate system. This measure is independent of the choice of the q's, and the unique measure in $\mathfrak{M}_{V^*}$ which agrees with it in each $\pi^{-1}(\mathcal{O})$ is the Liouville measure. We shall denote it by ζ. It can, of course, also be defined directly in an invariant manner starting from the fundamental tensor dW^0. It follows more or less immediately from its definition in invariant form that ζ is invariant under contact transformations. That is, $\zeta(E) = \zeta(A(E))$ whenever A is a contact transformation. In particular if $t \to U_t$ is the dynamical group of our system then $\zeta(U_t(E)) = \zeta(E)$ for all t and E.

Let ρ be any non-negative Borel function on $\mathfrak{M}_{V^*}$ such that $\int_{\mathfrak{M}_{V^*}} \rho \, d\zeta = 1$. Let $\alpha_\rho(E) = \int_E \rho \, d\zeta$. Then α_ρ will be a probability measure on $\mathfrak{M}_{V^*}$ and hence it will define a state. States of this form will be said to be states with a probability density—ρ being the probability density. There are cogent reasons for regarding them as having "greater physical reality" than other states. It follows at once from Liouville's theorem (i.e., the invariance of ζ) that $\hat{U}_t(\alpha)$ has a probability density for all t whenever α does. Indeed a straightforward calculation shows that $\hat{U}_t(\alpha_\rho) = \alpha_{\rho'}$, where $\rho'(q) = \rho(U_t^{-1}(q))$. If ρ_0 is differentiable and we set $\rho(t, q_1, \ldots, q_n, p_1, \ldots, p_n) = \rho_0(U_t^{-1}(q_1, \ldots, q_n, p_1, \ldots, p_n))$ then ρ satisfies the partial differential equation

$$\frac{\partial \rho}{\partial t} = -\frac{\partial \rho}{\partial q_1}\frac{\partial H}{\partial p_1} - \cdots - \frac{\partial \rho}{\partial q_n}\frac{\partial H}{\partial p_n} + \frac{\partial \rho}{\partial p_1}\frac{\partial H}{\partial q_1} + \cdots + \frac{\partial \rho}{\partial p_n}\frac{\partial H}{\partial q_n}$$

The first-order differential operator on the right is the infinitesimal generator of the one-parameter group of transformations on the functions ρ which $\hat{U}$ defines. The equation is the analog for classical statistical mechanics of the Schrödinger equation of quantum mechanics.

We shall be especially interested in those states that are "stationary" in the sense that they do not change with time. If ρ is an integral, for example, as well as a probability density then α_ρ will be a stationary state. We introduce now a one-parameter family of states called the *Gibbs canonical ensemble* and which is fundamental in the statistical interpretation of the theory of heat and thermodynamics. We are led to this family of states by asking the following question: Which state α_ρ represents the maximum amount of ignorance we can have about our system when we know the expected value of the energy, that is, when we know the value of the integral $\int \rho H \, d\zeta$? This question only makes sense when we have some way of measuring degree of randomness. The way to do this is suggested by information theory where arguments are given justifying the choice of $\alpha_1 \log 1/\alpha_1 + \cdots + \alpha_n \log 1/\alpha_n$ as a measure of the degree of ignorance represented by a set of n possibilities, the j-th of which occurs with probability α_j. Similar arguments justify the use of the analogous expression $\int \rho \log 1/\rho \, d\zeta$ for our purposes once one has accepted the perhaps somewhat dubious proposition that the (infinite) measure ζ represents complete and utter ignorance. At any rate we shall follow the traditional procedure (which is justified by its results) and adopt the indicated integral as a measure of randomness.

Using the standard techniques of the calculus of variations one finds that if ρ is to maximize $\int \rho \log 1/\rho \, d\zeta$ subject to the side condition $\int \rho H \, d\zeta = E$, where E is given, then ρ must be of the form $Ae^{-H/B}$, where A and B are constants. Of course since α_ρ is to be a probability measure we must have $A \int e^{-H/B} \, d\zeta = 1$, so that ρ must be of the form $e^{-H/B} / \int e^{-H/B} \, d\zeta$, where B is chosen so as to satisfy the equation

$$E = \frac{\int H e^{-H/B} \, d\zeta}{\int e^{-H/B} \, d\zeta}$$

In the systems which will be of interest to us in the present connection H will have the following properties:

(a) $H \geq 0$.

(b) $P(B) = \int e^{-H/B} \, d\zeta$ exists for all $B > 0$.

(c) $\int H e^{-H/B} \, d\zeta$ exists for all $B > 0$.

(d) $\int H e^{-H/B} \, d\zeta / P(B)$ is an unbounded non-decreasing function of B.

Under these assumptions, if we set

$$E(B) = \frac{\int H e^{-H/B} \, d\zeta}{P(B)}$$

we see that for each E there is just one B such that $\rho = e^{-H/B}/P(B)$ satisfies the required conditions. We shall call it the ***Gibbs canonical state*** for the given value of E. The value of $\int \rho \log 1/\rho$ when $\rho = e^{-H/B}/P(B)$ turns out to be $\log P(B) + E(B)/B$. We can compute the real-valued functions $P(B)$, $E(B)$, and $S(B) = \log P(B) + E(B)/B$ whenever our system satisfies (a), (b), (c), and (d). It is natural to ask for the physical significance of these functions and especially of the parameter B.

Suppose we take two independent systems with Hamiltonians H_1 and H_2 in the states defined by B_1 and B_2, respectively. Then the combined system will be in the state defined by the density $\exp[-(H_1/B_1 + H_2/B_2)]/P_1(B_1)P_2(B_2)$ and this will be a Gibbs canonical state if and only if $B_1 = B_2$. If $B_2 > B_1$ then the unique Gibbs state $\exp[-(H_1 + H_2/B)]/P(B)$, in which $E(B) = E_1(B_1) + E_2(B_2)$, will be defined by a B with $B_2 > B > B_1$ and we will have $E_1(B) > E_1(B_1)$ and $E_2(B) < E_2(B_2)$. Thus, to get a Gibbs state, energy must flow from the second system to the first until the B's are equal. In this respect the relationship between B and $E(B)$ is reminiscent of that between the *temperature* and the *heat* content of a material body.

Let us pause briefly to outline the main facts about the notions of heat and temperature. One's crude sense of hot and cold may be made precise by using the fact that most substances expand when heated. Arbitrarily chosing some substance (such as mercury) as a standard, one can measure temperature differences by measuring the expansion and contraction of a sample of the substance in a "thermometer." While the scale so found will depend upon the nature of the substance chosen, this is a temporary difficulty which we shall show below how to remove. When substances at different temperatures are put in contact their temperatures change until they arrive at some common intermediate value. This intermediate value will depend upon the relative amounts of the substance present being nearer to the larger value, the more there is of the hotter substance present, and vice versa. Exact measurements suggest and permit introducing a more or less exact notion of *quantity of heat*—unit quantity of heat being the amount of heat that is required to raise a standard amount of a standard substance through a standard temperature difference. When heat passes from one body to another body the amount of heat gained by one is just equal to the amount lost by the other. Note that if we measure quantity of heat in terms of the amount of standard substance that can be raised through the standard temperature interval we do not have to face the fact that our arbitrary temperature scale provides only a rather dubious way of deciding when two temperature differences are equal.

The above remarks about conservation of quantity of heat only apply insofar as the potential energy of all mechanical systems remains unchanged in the expansion and contraction that takes place. For example a gas that lifts a weight as it expands will cool without heating up anything else. On the other hand, when energy disappears from an otherwise conservative system due to "friction" the material bodies concerned always rise in temperature without cooling anything else. One of the great discoveries of the mid-nineteenth century was that there is a quantitative conservation law connecting heat and mechanical energy. When energy is dissipated by friction, the amount of heat that appears is directly proportional to the amount of energy that disappears. Moreover the proportionality constant is the same as that occurring when heat is converted into mechanical energy by expansion. This discovery of course greatly strengthened the position of those who held a rise in temperature to be due to an increase in the invisible motions of the atoms of a substance, and this view came to be universally accepted.

Although heat is a form of mechanical energy, it is a relatively "unavailable" form and is not *freely* convertible into the latter. One can convert heat energy into mechanical energy only by "paying for it" in some way and, in particular, it is impossible to take a system through a closed cycle of changes the only effect of which is to convert a certain amount of heat energy into mechanical energy. This principle can be precisely formulated in a number of equivalent ways which are collectively known as the *second law of thermodynamics*. One form asserts that if a "heat engine" works "reversibly" between two temperatures absorbing heat at the higher, T'_1, giving some of it up at the lower, T'_2, and converts a fraction f of it into mechanical energy, then f is independent of the nature of the engine and depends only upon the temperatures involved. It is easy to deduce the existence of a monotone function ϕ unique up to a multiplicative constant such that $f(T'_1, T'_2) = [\phi(T'_1) - \phi(T'_2)]/\phi(T'_1)$. Setting $T = \phi(T')$ we get a natural *absolute* temperature scale independent (except for the normalizing constant) of the properties of any particular substance and such that $f(T_1, T_2) = T_1 - T_2/T_1$. It is customary to normalize ϕ so that there are 100 temperature units of difference (100 degrees) between the freezing and boiling points of water.

The invariance of f has strict quantitative implications concerning the connections between the expandibility of material substances and their capacity to absorb heat—any departure from which could be exploited to build a so-called "perpetual motion machine of the second kind." These take the form that a certain differential defined on the manifold of "equilibrium states" is the differential of a function. This function, unique up to an additive constant, is called the "entropy" and plays a vital role in the theory.

In the later part of the nineteenth century attempts were made to deduce the second law of thermodynamics and its consequences from the laws of mechanics, and the hypothesis that heat energy is the ordinary mechanical energy of the atoms of which the body is made. These attempts culminated in the statistical mechanics of Gibbs and the doctrine that the unavailability of heat energy is quantitatively related to the degree of our ignorance of the exact motions producing it. In precise terms it was found that one obtained a model for thermodynamical phenomena by representing the state of the atomic motions by the Gibbs canonical ensemble and interpreting B as absolute temperature, $E(B)$ as heat content, and $S(B)$ as entropy. Of course, B being in energy units and temperature in degrees, a conversion factor is necessary. The corresponding "universal constant" is called Boltzmann's constant and is usually denoted by k. $B = kT$.

Any hypothesis about the structure of matter which represents it as a system of interacting particles having a Hamiltonian satisfying the conditions (a), (b), (c), and (d) laid down above yields, via the Gibbs state, formulas $T \to E(kT)$ and $T \to S(kT)$ for heat energy and entropy as a function of absolute temperature which may be compared with the results of experiment.

We conclude this section with a study of certain special cases which are important in making the transition to quantum theory. First, however, we make some general remarks about the technique of computing P, B, and S. To begin with we do not really need to integrate over $\mathfrak{M}_V{}^*$. All operations can be transferred to the real line in the following way. For each Borel subset F of the real line let $\beta(F) = \zeta(H^{-1}(F))$. It is then an easy exercise in real variable theory to show that

$$P(B) = \int_0^\infty e^{-x/B}\, d\beta(x)$$

and

$$E(B) = \frac{\int_0^\infty x\, e^{-x/B}\, d\beta(x)}{P(B)}$$

where β is the measure on the real line just defined. Moreover applying the standard theory of the Laplace transform we may differentiate $P(B)$ by Leibniz's rule and conclude that

$$P'(B) = \int_0^\infty e^{-x/B}\, x/B^2\, d\beta(x) = 1/B^2 \int_0^\infty x\, e^{-x/B}\, d\beta(x)$$

$$= \frac{E(B)\,P(B)}{B^2}$$

Thus $E(B) = B^2\, d/dB \log P(B)$ and we have only one integral to compute since $S(B) = \log P(B) + (E(B)/B)$. The function $P(B)$ from which everything else may be computed is called the "partition function." By the theory of the Laplace transform P and B determine one another uniquely.

As a first example let us compute β, P, E, and S for a "perfect gas," the latter being conceived of as a set of n mass points in an enclosure of volume V in such a way as not to interact at all. H will then be $p_1^2/2M_1 + \cdots + p_{3n}^2/2M_{3n}$ whenever the q's are such that all particles are in the enclosure and will be ∞ whenever the q's are such that one particle is not in the enclosure. It is an easy exercise in higher dimensional integration to compute that

$$\beta([0, x]) = V^n \sqrt{M_1 \cdots M_{3n}}\ C_n\, x^{3n/2}$$

where C_n depends only on n. Let $D_n = C_n\sqrt{M_1 \cdot \cdots \cdot M_{3n}}$. Then

$$P(B) = 3n/2\ D_n\, V^n \int_0^\infty e^{-x/B}\, x^{3n/2\,-1}\, dx$$

By setting $x = yB$ we get

$$\begin{aligned} P(B) &= 3n/2\, D_n\, V^n \int_0^\infty e^{-y}\, y^{3n/2\,-1}\, B^{3n/2}\, dy \\ &= 3n/2\, D_n\, V^n\, B^{3n/2} \int_0^\infty e^{-y}\, y^{3n/2\,-1}\, dy \\ &= V^n\, B^{3n/2}\, A_n \end{aligned}$$

where A_n depends only on n and the M_j. Finally

$$\begin{aligned} E(B) &= B^2 \frac{d}{dB} \log P(B) = B^2 \frac{d}{dB}[\log A_n + n \log V + 3n/2 \log B] \\ &= B^2\ 3n/2\ \ 1/B = 3/2\ \ nB \end{aligned}$$

Thus $E(kT) = 3/2\ nkT$ and we have the classical result that the specific heat [i.e., $d/dt\ E(kT)$] of a perfect monatomic gas is independent of the volume and temperature and is directly proportional to the number of atoms present. By measuring $d/dt\ E(kT)$ experimentally one can find nK and thus deduce either k or n (the number of atoms in a given weight of a gas) when the other is known. We have also

$$\begin{aligned} k\, S(kT) &= k \log A_n + nk \log V + 3/2\ nk \log kT + 3/2\ nk \\ &= nk \log V + 3/2\ nk \log T + A_n' \end{aligned}$$

where A_n' depends only on n. Classically entropy is defined only up to an additive constant and in its dependence on V and T the expression just derived is exactly that of the classical theory.

As a second example let us think of a solid as a system of n atoms interacting in such a way as to have a configuration of stable equilibrium about which it can vibrate according to the laws of small oscillations. We shall compute $E(kT)$ for such a system and derive a formula for the temperature dependence of the specific heat of a solid.

As we have seen, coordinates may be introduced for a linear system in such a way that it decomposes into independent systems having one-dimensional configuration spaces, the single coordinate in each of these spaces undergoing simple harmonic oscillations. On the other hand, it is easy to show that the P for a product of two independent systems is the product of the corresponding P's. Thus if P^{ν} denotes P for a simple harmonic oscillator of frequency ν and the frequencies of the simple harmonic oscillations into which the motion of our system decomposes are $\nu_1, \ldots, \nu_{3n}$, then P for our system will be such that

$$P(B) = P^{\nu_1}(B) \cdot \ldots \cdot P^{\nu_{3n}}(B)$$

Now the Hamiltonian of a harmonic oscillator is of the form $(p^2/2M) + \mu/2\, q^2 = H(q, p)$. Thus $\beta([0, x])$ is the area of $(p^2/2M) + \mu/2\, q^2 \leq x$, i.e., the area inside the ellipse

$$\frac{p^2}{(\sqrt{2Mx})^2} + \frac{q^2}{(\sqrt{2x/\mu})^2} = 1$$

i.e., $\pi(\sqrt{2Mx})(\sqrt{2x/\mu}) = 2\pi x\sqrt{M/\mu} = x/\nu$, where ν is the frequency of the oscillator. In other words β is simply $1/\nu$ times Lebesgue measure in the line. We compute at once that

$$P^{\nu}(B) = 1/\nu \int_0^{\infty} e^{-x/B}\, dx = B/\nu$$

Hence

$$P(B) = \frac{B^{3n}}{\nu_1 \cdot \ldots \cdot \nu_{3n}}$$

Thus

$$E(B) = B^2 \frac{d}{dB}\left[\log B^{3n} - \log(\nu_1 \cdot \ldots \cdot \nu_{3n})\right] = 3nB$$

so $E(kT) = 3nkT$. We deduce that the specific heat is independent of the characteristic frequencies and of the temperature and depends only upon the number of atoms present. Now an empirical law discovered in 1819 by Dulong and Petit asserted that all elements in solid form have the same *atomic* specific heat, i.e., the same specific heat when allowance is made for different numbers of atoms per unit weight (i.e., different atomic weights). This "law" turned out to have many exceptions but to be pretty generally true at sufficiently high temperatures. Statistical mechanics provides an explanation for the law but not for its failure at low temperatures.

The fact that each independent oscillation makes the same contribution to $E(B)$ regardless of its frequency is called the *law of equipartition of energy*. When one attempts to apply it to systems with an infinite number of degrees of freedom in order to study the influence of temperature on radiation one is led to certain serious anomalies. In the first section of the next chapter we shall present these anomalies and explain how their partial elucidation led to the beginnings of the quantum theory.

Chapter 2

QUANTUM MECHANICS

2-1 The Old Quantum Theory

At the end of the nineteenth century physicists had good reason to feel self-satisfied. The plan of reducing all phenomena to Newtonian mechanics had gone on from success to success and now encompassed light, electromagnetics, and heat in a very satisfying way. Indeed it had been conjectured that physics was essentially complete—that little remained to be done but find the next decimal point in the fundamental constants. However there were clouds upon the horizon and experiments were underway which were to upset the traditional scheme pretty completely.

Although one thought of macroscopic phenomena as ultimately due to the motions of submicroscopic particles, it was only toward the end of the nineteenth century that it became possible to study these particles as individuals—and individually the particles were remarkably non-Newtonian. It is interesting to note that Maxwell, who was a leader both in the development of electromagnetic theory and in the statistical theory of heat, was also the organizer and first director of the Cavendish laboratory in Cambridge, England, where many of the most fundamental elementary particle experiments were made. Maxwell headed this laboratory from its founding in 1871 to his death in 1879. He was succeeded for a 5-year interim period by Lord Rayleigh and in 1884 the 28-year-old J. J. Thomson took over. Thomson remained in the post until succeeded in 1919 by his student Rutherford. It was in this laboratory that Thomson discovered the electron. C. T. R. Wilson invented the cloud chamber, and much other important progress was made.

Although it is on the atomic level that the failure of Newtonian physics is most complete, there are macroscopic phenomena in which this failure makes itself felt, and it was in the attempt to explain these that quantum theory began.

Consider the electromagnetic field inside a finite enclosure. As we have seen, it may be regarded as a linear mechanical system and hence as an assembly of independent oscillators. As such, in any given state there will be a certain distribution of its energy among these oscillators and this distribution may be measured experimentally by the techniques of spectroscopy. Applying the notions of thermodynamics one expects Gibbs states of "thermal equilibrium," one for each temperature, and one can observe these experimentally. Under suitable experimental conditions there is a spectral distribution of energy in enclosed radiation which depends on the temperature of the enclosure but not on the material of which the enclosure is made. On the other hand, attempts to explain the experimental distribution theoretically led to paradoxical results in a manner which we shall now explain.

There is a difficulty right at the start arising from the fact that we are dealing with a system with an *infinite* number of degrees of freedom. This means, on the one hand, that we have no Liouville measure from which to construct a Gibbs state, and, on the other, that when we analyze the system into independent harmonic oscillations we get a countable infinity of them with frequencies $\nu_1, \nu_2, \ldots$ tending to ∞. One can attempt to circumvent the first difficulty by applying the results of an analysis of finite dimensional, linear systems directly to the infinite dimensional case. In 1900 Rayleigh did just this. He assumed that the equipartition law continued to hold and, ignoring the fact that this implies an infinite total energy, derived a spectral distribution law. If the enclosure is rectangular the fundamental frequencies $\nu_1, \nu_2, \ldots$ may be obtained by a straightforward application of Fourier analysis, and it is found that for a sufficiently large volume V one has approximately $(8\pi\nu^2/c^3)\,V\Delta\nu$ ν_j's between ν and $\nu+\Delta\nu$. Here c is the velocity of light and $\Delta\nu$ is small compared to ν but large enough to include a representative number of ν_j's. The larger ν is, the closer one is to a continuous distribution with density $8\pi\nu^2/c^3$. Applying the equipartition law we assign an energy kT to each ν_j and arrive at *Rayleigh's formula*, according to which there is an energy per unit volume of $8\pi\nu^2 kT\Delta\nu/c^3$ in the frequency range between ν and $\nu+\Delta\nu$ for radiation in "thermal equilibrium" at temperature T. Rayleigh was rewarded for his boldness by finding that his formula was in agreement with experiment for low frequencies and high temperatures —more precisely, when ν/T is sufficiently small. As might be expected, it was way out of line for ν/T large—experiments did not suggest an infinite total energy. However there was another law, derived a few years earlier, by Wien, which agreed with experiment for large ν/T. We shall not attempt to describe Wien's rather involved physical reasoning, but simply announce that he replaced Rayleigh's formula by $A\nu^3 e^{-b\nu/T}\Delta\nu$, where A and b are constants which were left to be determined by the experimental data. This formula denies equipartition and does not fit the experimental facts for small ν/T. On the other hand, it avoids the paradox of infinite total energy. Reconciliation of the complementary and contradictory laws of Rayleigh and Wien presented a major challenge to the physicists of 1900—a challenge which was successfully met by Max Planck.

We shall now derive Planck's law by arguments analogous to, but rather different from, those actually used by Planck.

Suppose that in deriving P from β in the case of the harmonic oscillator we replace β by a discrete measure that approximates it. We recall that β is $1/\nu$ times Lebesgue measure. It may be approximated by a measure β_h concentrated in equally spaced points and giving the same small measure h to each point. If the points are a units apart then the interval $0 \leq t \leq Na = x$ will have measure Nh. If this is equal to x/ν we must have $Na/\nu = Nh$ or $a = h\nu$. We now compute:

$$P_h(B) = \int_0^\infty e^{-x/B}\, d\beta_h(x)$$

$$= h + he^{-a/B} + he^{-2a/B} + \cdots$$

$$= h(1 + e^{-h\nu/B} + e^{-2h\nu/B} + \cdots)$$

We have a geometric series with ratio $e^{-h\nu/B}$ whose sum is $h/(1 - e^{-h\nu/B})$. The corresponding E is given by

$$E_h(B) = B^2 \frac{P_h'(B)}{P_h(B)} = \frac{h\nu e^{-h\nu/B}}{1 - e^{-h\nu/B}} = \frac{h\nu}{e^{h\nu/B} - 1}$$

For a system which may be analyzed into independent oscillators with frequencies $\nu_1, \nu_2, \ldots$ we get the *convergent* expression:

$$E_h(B) = \frac{h\nu_1}{e^{h\nu_1/B} - 1} + \frac{h\nu_2}{e^{h\nu_2/B} - 1} + \cdots$$

and Rayleigh's formula becomes

$$\frac{8\pi\nu^3}{c^3} \frac{h}{e^{h\nu/kT} - 1} \Delta\nu$$

Of course, as $h \to 0$, $E_h(B) \to \infty$ and the new Rayleigh's formula goes over into the old one. On the other hand, no matter what value we give to h, if we take ν/T so small that $e^{h\nu/kT}$ may be approximated by the first two terms of the Taylor series then $e^{h\nu/kT} - 1$ becomes $h\nu/kT$ and again we recover the old Rayleigh formula. Moreover, if we choose ν/T so large that 1 may be neglected in comparison with $e^{h\nu/kT}$, then the formula becomes

$$\frac{8\pi\nu^3}{c^3}\ h\, e^{-(h/k)(\nu/T)}$$

which is identical with Wien's if we let $A = 8\pi h/c^3$ and $b = h/k$. Indeed these two equations determine h and k uniquely, and the value of k so found agrees with that found by other means. Planck originally meant to let h tend to zero but found that he got a unified formula by stopping short at the value determined by the above equation. The new physical constant so defined now bears his name, as does the resulting radiation law.

As we have seen, Planck's law contains those of Wien and Rayleigh as special cases. It also agrees with experiment over the entire frequency range. Its only drawback is the arbitrary nature of the decision to replace β by β_h, which amounts to the rather mystical assumption that a harmonic oscillator of frequency ν can only have energies that are multiples of $h\nu$.

Planck's device can also be used to obtain more accurate specific heat formulas. If we replace kT by $h\nu/(e^{h\nu/kT} - 1)$ in the formula for the energy content of a solid, we get

$$E(kT) = \sum_{j=1}^{3n} \frac{h\nu_j}{e^{h\nu_j/kT} - 1}$$

instead of 3nkT. Far from being independent of the ν_j the specific heat depends upon them in a rather complicated way. When sufficient work had been done on the problem of estimating the ν_j, quite accurate formulas were obtained. In any event, when $h\nu_j/kT$ is much less than one for all ν_j, we see that E(kT) is approximately 3nkT. Thus we understand why the law of Dulong and Petit holds for high temperatures but not for low ones. This application to specific heats was begun by Einstein in 1906 and continued by others, notably P. Debye.

In 1905 Einstein showed how Planck's ideas could be extended to give a quantitative explanation of the so-called "photoelectric effect" discovered in 1899 by J. J. Thomson and P. Lenard. These physicists (working independently of one another) found that ultraviolet light falling on a metal plate causes it to emit electrons. Moreover, in 1902 Lenard had shown that these electrons have an initial energy quite independent of the intensity of the light and depends only upon its frequency. According to the classical theory of charges interacting with the electromagnetic field, the energy of the electrons should depend strongly on the intensity of the light and with weak light there should be a long time lag before an electron accumulates enough energy to escape. The time lag is not observed. Einstein pointed out that if one supposes that the energy in an electromagnetic field with fundamental frequencies $\nu_1, \nu_2, \ldots$ occurs in "packets" with energies $h\nu_1, h\nu_2, \ldots$ as is suggested by Planck's radiation formula, then one can explain the photoelectric effect by supposing that these packets collide with an electron and give up their energy to it. The quantitative implications of this hypothesis were accurately verified by a number of experiments and the long dead corpuscular theory of light was revived.

One of the most striking applications of "quantum" ideas was made by N. Bohr in 1913, who used them to deduce a theoretical formula for the lines in the spectrum of hydrogen. Every chemical element, when suitably excited, emits radiation whose spectrum, when analyzed, contains certain *sharp lines* characteristic of the element in question. An obvious problem was that of correlating these lines with the internal structure of the atoms of the element concerned. However not enough was known about this internal structure to make the problem a hopeful one until Rutherford made his fundamental discovery of 1911. Working in Manchester on the scattering by matter of the "α particles" emitted by radioactive substances, Rutherford came to the conclusion that an atom must consist of a very small positively charged nucleus surrounded by enough electrons to balance the charge. These electrons, he decided, must be distributed throughout a value of diameter about 300 times that of the nucleus. He supposed further that the number of electrons would be about half the atomic weight (except for hydrogen, where this would be impossible) and that this number determined the chemical properties of the element in question. This model, with later refinements and adapted to quantum mechanics, is now universally accepted. Applying classical mechanics to Rutherford's model for the (one-electron) hydrogen atom and ignoring electromagnetic theory, except for electrostatics, one finds that when the energy lies below a certain limit the electron will travel in an elliptical orbit about the center of mass of itself and the nucleus. In each orbit the electron will have a constant energy, but as we go from orbit to orbit a continuum of energies is possible. Bohr found that he could predict the principal features of the spectrum of hydrogen by making the following assumptions:

(a) Only those orbits are possible in which the angular momentum is a multiple of $h/2\pi$.

(b) When an atom shifts from one possible orbit to another the energy difference, $E_1 - E_2$ is transformed into radiation of frequency $E_1 - E_2/h$, i.e., into a "photon" or packet of electromagnetic energy of energy $E_1 - E_2 = h\nu$.

Bohr's idea was later extended so as to apply to more complex atoms and in a short time the complex data of spectroscopy was organized and understood in a manner which had previously seemed hopelessly unattainable.

In spite of its many successes this device of combining the laws of classical mechanics with *ad hoc* "quantization rules" was obviously unsatisfactory. Not only was there no theoretical foundation for the rules but also no general way of deriving the rules. New complications in the system being studied led to ever more elaborate rules for deciding which discrete values of the variable were to be allowed. Moreover in many ways the new rules were in flat contradiction to classical physics and yet did not completely replace it. In explaining some phenomena one used the wave theory of light and in explaining others the new corpuscle theory. Again, according to classical electrodynamics the electron in a Bohr orbit should have radiated away all its energy and fallen into the nucleus, but this did not happen.

Finally in 1925 Heisenberg and Schrödinger independently discovered general quantization rules which were superficially rather different but were soon seen to be equivalent. Their discoveries set the world of physics on fire and in half a decade the combined efforts of Dirac, Bohr, Born, von Neumann, the two named above, and others had produced a new and more fundamental mechanics which:

(a) Had classical mechanics as a limiting case for large masses and distances.

(b) Implied the existence and exact form of the quantization rules.

(c) Made clear exactly how matter and radiation were at the same time waves and particles and clarified the stability of the Bohr orbits.

(d) Explained many hitherto inexplicable phenomena, e.g., valence and the mechanism of the formation of chemical compounds.

This new mechanics is called "quantum mechanics" and it is to its development that we now turn. We should like to emphasize that in quantum mechanics the quantum rules are a deduction and not a fundamental feature of the formulation of the theory. In this sense the name quantum mechanics is unfortunate. It is basically a revision of statistical mechanics in that one studies the change in time of probability measures but no longer supposes that the motion of these measures is that induced by a motion of points in phase space. The laws of quantum mechanics place certain restrictions on the possible *simultaneous* probability distributions of various observables and give differential equations which may be integrated to show how they change in time. Everything else is deducible from these laws.

2-2 The Quantum-Mechanical Substitute for Phase Space

In classical statistical mechanics an observable is a Borel function on the phase space $\mathcal{M}_V*$, a state is a probability measure on $\mathcal{M}_V*$, and to each pair consisting of an observable f and a state α we have associated the probability measure α_f on the real line which takes the Borel set E into $\alpha(f^{-1}(E))$. The number $\alpha(f^{-1}(E))$ = (definition) $p(f, \alpha, E)$ is the probability that a measurement of f will be in E when the system is in the state α.

In quantum mechanics we shall also have a function p assigning a probability to each triple consisting of an observable A, a state α, and a Borel set E of real numbers. However, we shall not suppose that p may be derived from a probability measure in classical phase space.

The development of quantum mechanics led to a close examination of the measurement process and this, in turn, led to the doctrine (known when formulated in quantitative terms as the Heisenberg uncertainty principle) that it is in principle meaningless to speak of *exact* simultaneous values for the position and velocity coordinates of a particle. Position and velocity measurements interfere with one another in such a fashion as to frustrate all attempts to assign a precise operational meaning to the statement: At the time t the particle P had position coordinates x, y, z and velocity components v_x, v_y, v_z. On the other hand it *is* possible to assign such a meaning to the statement: At the time t the state of the system was such that measurements of x, y, z, v_x, v_y, v_z were statistically distributed according to the real line probability measures $\alpha_1, \alpha_2, \alpha_3, \beta_1, \beta_2, \beta_3$. This is because probability distributions are determined by making measurements on a large number of replicas of a given system and we can use a different set of samples for each observable. The process of making the measurement, of course, changes the state and the information we get applies only to the unexamined members of our family of replicas. For a full discussion of this important point the reader is referred to the book of von Neumann.

Since a point in phase space has no physical meaning, neither does the notion of a probability measure in phase space. On the other hand, the notion of the assignment of a probability measure on the line to each observable does, as we have seen, have a meaning. We take this notion as the quantum-mechanical refinement of the notion of state in classical statistical mechanics.

It will be convenient to proceed axiomatically. We shall build up a rigorously defined mathematical model and describe its physical meaning as precisely as we can. As the reader will see the axioms will have varying degrees of physical plausibility. It is not yet possible to deduce the present form of quantum mechanics from completely plausible and natural axioms.

Let $\mathcal{B}$ be the set of all Borel subsets of the real line R. We suppose we are given two abstract sets $\mathcal{O}$ and $\mathcal{S}$ and a function p which assigns a real number $p(A, \alpha, E)$ in $0 \leq x \leq 1$ to each triple A, α, E, where A is in $\mathcal{O}$, α is in $\mathcal{S}$, and E is in $\mathcal{B}$. We assume that p has certain properties which we list as axioms. Physically $\mathcal{O}$ is to be thought of as the set of all observables of our system, and $\mathcal{S}$ as the set of all states. $p(A, \alpha, E)$ is the probability that a measurement in the state α of A will lead to a value in E. We think now of time as standing still. Later on we shall discuss the change of states as time elapses.

Axiom I: $p(A, \alpha, \phi) = 0, p(A, \alpha, R) = 1$ (ϕ represents the empty set)

$$p(A, \alpha, E_1 \cup E_2 \cup \ldots) = \sum_{j=1}^{\infty} p(A, \alpha, E_j)$$

whenever the E_j are Borel sets that are disjoint in pairs. Let us set $\alpha_A(E) = p(A, \alpha, E)$. Axiom I simply states that for each A in $\mathcal{O}$ and each α in $\mathcal{S}$, α_A is a probability measure.

Axiom II: If $p(A, \alpha, E) = p(A', \alpha, E)$ for all α and E then $A = A'$. Similarly if $p(A, \alpha, E) = p(A, \alpha', E)$ for all A and E, then $\alpha = \alpha'$. Axiom II says that two states, to be different, must assign different probability distributions to at least one observable and that two observables, to be different, must have different probability distributions in at least one state.

Axiom III: Let A be any member of $\mathcal{O}$ and let f be any real-valued Borel function on the line. Then there exists B in $\mathcal{O}$ such that $p(B, \alpha, E) = p(A, \alpha, f^{-1}(E))$ for all α in $\mathcal{S}$ and all E in $\mathcal{B}$.

It follows from Axiom II that B is uniquely determined by A and we shall denote it by $f(A)$.

Physically the observable $f(A)$ is constructed from A as follows. Whatever we do to measure A we measure $f(A)$ by applying the function f to the result of measuring A. If we think of what this does to probabilities, we are led to Axiom III.

Because of Axiom III, expressions such as $A^2, A^3 + A$, $1 - A$, and e^A all make sense whenever A is an observable. In particular if $f(x) \equiv \lambda$, where λ is some real number, then $f(A)$ is independent of A and will be called the constant observable with value λ or simply the observable λ.

Axiom IV: If $\alpha_1, \alpha_2, \ldots$ are members of $\mathcal{S}$ and $t_1 + t_2 + \cdots = 1$, where $0 < t_i < 1$, then there exists α such that

$$p(A, \alpha, E) = \sum_{j=1}^{\infty} t_j \, p(A, \alpha_j, E)$$

for all E in $\mathcal{B}$ and all A in $\mathcal{O}$.

It follows from Axiom II that α is uniquely determined by the α_j and t_j. We denote it by

$$\sum_{j=1}^{\infty} t_j \, \alpha_j$$

It corresponds physically to a state in which we know that we are in the state α_i with probability t_i. It is said to be a *mixture* of the states α_j. If α cannot be obtained by mixing two states different from itself it is said to be a *pure state*. [We warn the physicist reader that Axiom IV is *not* a formulation of the "superposition principle." The latter yields pure states while Axiom IV yields mixed ones.]

Before listing any more axioms it will be useful to introduce some notions based on those we have already. Let us call an observable A, a *question*, if in every state α the measure α_A is concentrated in the points 0 and 1, that is, if $\alpha_A(\{0, 1\}) = 1$ for all α. We leave it to the reader to verify that A is a question if and only if $A^2 = A$. We shall denote the set of all questions by the symbol $\mathcal{Q}$. That questions exist in great profusion may be seen as follows. Let E be any Borel subset of the real line and let ϕ_E be the function which is 1 for x in E and 0 when x is not in E. Then $\phi_E(A)$ is seen at once to be a question. We shall denote this question by Q_E^A. It is the observable which yields the value 1 whenever a measurement of A yields a value in E, and the value 0 otherwise. In this sense it corresponds to asking the (yes or no) question: Did the measurement of A lead to a value in E?

If we hold A fixed, then Q_E^A is a family of questions parameterized by the Borel sets E. It is important for our purposes to note that this parameterized family of questions determines A uniquely. That is, as the reader may easily prove for himself, $Q_E^A = Q_E^{A'}$ for all E in $\mathcal{B}$ implies $A = A'$. Let Q be any question and let α be any state. Let $\alpha_Q(\{1\}) = s$. Then $\alpha_Q(\{0\}) = 1 - s$, and for any set E in $\mathcal{B}$, $\alpha_Q(E)$ is $0, 1, s$, or $1 - s$, according as neither 0 nor 1 is in E, both 0 and 1 are in E, 1 is in E but 0 is not, 0 is in E and 1 is not. Thus α_Q is completely determined by $s = \alpha_Q(\{1\})$. We shall define $m_\alpha(Q)$ as $\alpha_Q(\{1\})$. Then m_α is a certain real-valued function on the questions and the reader should have no trouble in verifying that $m_{\alpha_1}(Q) = m_{\alpha_2}(Q)$ for all Q implies $\alpha_1 = \alpha_2$.

The functions m_α define a natural partial ordering in $\mathcal{Q}$ as follows: $Q_1 \leq Q_2$ if and only if $m_\alpha(Q_1) \leq m_\alpha(Q_2)$ for all states α. It is easy to check that the relation so defined satisfies the defining conditions of a partial ordering, that is:

(1) $Q_1 \leq Q_2$ and $Q_2 \leq Q_3$ implies $Q_1 \leq Q_3$.
(2) $Q \leq Q$ for all Q.
(3) $Q_1 \leq Q_2$ and $Q_2 \leq Q_1$ implies $Q_1 = Q_2$.

For any question Q the observable $1 - Q$ is also a question. It is the question whose answer is "yes" if and only if the answer to Q is "no." Q and $1 - Q$, unlike pairs of questions in general, are both functions of the same observable, namely Q, and hence can be asked simultaneously.

Let Q_1 and Q_2 be questions. If $Q_1 \leq 1 - Q_2$, or, equivalently, if $m_\alpha(Q_1) + m_\alpha(Q_2) \leq 1$ for all states α, we shall say that Q_1 and Q_2 are disjoint and write $Q_1 \delta Q_2$. Physically, $Q_1 \delta Q_2$ means that Q_1 and Q_2 cannot have simultaneous yes answers.

To understand further the meaning of $Q_1 \leq Q_2$ and $Q_1 \,\delta\, Q_2$ consider the case in which $Q_1 = Q^A_{E_1}$ and $Q_2 = Q^A_{E_2}$, where A is some observable. Then $E_1 \subset E_2$ implies $Q^A_{E_1} \leq Q^A_{E_2}$ and $E_1 \cap E_2 = \phi$ implies $Q^A_{E_1} \,\delta\, Q^A_{E_2}$.

We pause at this stage to point out that all our axioms so far are satisfied for the p of classical mechanics and to see what our various constructs amount to in the classical case. In this case an observable is simply a Borel function g on $\mathfrak{M}_{V^*}$. The reader can check without difficulty that $f(g)$ as defined in Axiom III is just the usual composite function $f \circ g$. A question is simply a Borel function that takes on only the values 0 and 1. It is uniquely determined by the Borel set on which it takes the value 1, and it is convenient to identify it with this set. If this is done, $Q \to 1 - Q$ becomes the operation of going from a set to its complement, $Q_1 \leq Q_2$ becomes set inclusion, and $Q_1 \,\delta\, Q_2$ becomes the disjointness of sets. The function m_α mapping questions into numbers becomes the measure α on $\mathfrak{M}_{V^*}$, and the function $E \to Q^g_E$, where g is a Borel function on $\mathfrak{M}_{V^*}$, becomes the mapping $E \to g^{-1}(E)$.

Returning to the general case let us note that if A is an observable and $E_1, E_2, \ldots$ are disjoint Borel sets with $E = E_1 \cup E_2 \cup \ldots$ then for every state α we have

$$m_\alpha(Q^A_E) = p(A, \alpha, E) = \sum_{j=1}^{\infty} p(A, \alpha, E_j) = \sum_{j=1}^{\infty} m_\alpha(Q^A_{E_j})$$

Moreover, by Axiom II, Q^A_E is uniquely determined by the fact that for all α,

$$m_\alpha(Q^A_E) = \sum_{j=1}^{\infty} m_\alpha(Q^A_{E_j})$$

It is natural to say that Q^A_E is the sum of the questions $Q^A_{E_1}, Q^A_{E_2}, \ldots$ and write $Q^A_E = Q^A_{E_1} + Q^A_{E_2} + \cdots$. Generalizing we shall say that the question Q is the sum of the disjoint questions $Q_1, Q_2, \ldots$ and write $Q = Q_1 + Q_2 + \cdots$ if for all α in $\mathcal{S}$ $m_\alpha(Q) = m_\alpha(Q_1) + m_\alpha(Q_2) + \cdots$. By Axiom II there can be at most one such Q for a given sequence $Q_1, Q_2, \ldots$; however there is no guarantee that any such Q exists. We take care of this with our next axiom.

Axiom V: Let $Q_1, Q_2, \ldots$ be any sequence of questions such that $Q_i \,\delta\, Q_j$ for $i \neq j$. Then $Q_1 + Q_2 + \cdots$ exists.

Q may be interpreted physically as the question whose answer is yes if and only if at least one Q_i has the answer yes. It is useful to note that $Q_1 + Q_2 + \cdots$ is also the least question that is greater than or equal to each Q_i, so that it may be defined in purely order-theoretical terms. This fact is not quite obvious and we give a proof (due to R. V. Kadison).

Theorem: If $R \geq Q_j$ for all j and $Q_i \,\delta\, Q_j$ for $i \neq j$, then $R \geq Q_1 + Q_2 + \cdots$.

Proof: We show first that if $Q' \,\delta\, Q''$, $R \,\delta\, Q'$, and $R \,\delta\, Q''$ then $R \,\delta\, (Q' + Q'')$. Indeed $R + Q' + Q''$ must also be a question. Denote it by Q'''. Then for all α, $m_\alpha(R) + m_\alpha(Q') + m_\alpha(Q'') = m_\alpha(Q''') \leq 1$. Hence $m_\alpha(R) + m_\alpha(Q' + Q'') \leq 1$. Hence $R \,\delta\, (Q' + Q'')$. Now since $R \geq Q_j$ for all j, $(1 - R) \,\delta\, Q_j$ for all j, so by induction $(1 - R) \,\delta\, (Q_1 + \cdots + Q_n)$ for all n. Hence $R \geq Q_1 + \cdots + Q_n$ for all n. Hence $m_\alpha(R) \geq m_\alpha(Q_1 + \cdots + Q_n) = m_\alpha(Q_1) + \cdots + m_\alpha(Q_n)$ for all n and α. Hence $m_\alpha(R) \geq m_\alpha(Q_1) + m_\alpha(Q_2) + \cdots$ for all α. Hence $R \geq Q_1 + Q_2 + \cdots$.

In terms of this notion of sum for disjoint questions we can write down certain important properties of the functions $E \to Q_E^A$ from $\mathcal{B}$ to $\mathcal{Q}$ and $Q \to m_\alpha(Q)$ from $\mathcal{Q}$ to the reals. For each A, Q^A has the properties:

(a) $E \cap F = \phi$ implies $Q_E^A \,\delta\, Q_F^A$.
(b) $E_i \cap E_j = \phi$ for $i \neq j$ implies $Q_{E_1 \cup E_2 \cup \cdots}^A = Q_{E_1}^A + Q_{E_2}^A + \cdots$.
(c) $Q_\phi^A = 0$, $Q_R^A = 1$.

If $q: E \to q_E$ is any function from $\mathcal{B}$ to $\mathcal{Q}$ that satisfies (a), (b), and (c) we shall call it a *question-valued measure*. It follows from a theorem proved earlier that each observable A is uniquely determined by the question-valued measure $E \to Q_E^A$ associated with it, so that we have a natural one-to-one correspondence between observables and *certain* question-valued measures. Our next axiom makes it possible to omit the word "certain."

Axiom VI: If q is any question-valued measure then there exists an observable A such that $Q_E^A = q_E$ for all E in $\mathcal{B}$.

The significance of this axiom may be explained as follows. Suppose first that it does not hold. Let $\mathcal{O}'$ denote the set of all question-valued measures. For each q in $\mathcal{O}'$, each α in $\mathcal{S}$, and each E in $\mathcal{B}$ let $p'(q, \alpha, E) = m_\alpha(q_E)$. It is easy to see that $\mathcal{O}'$, $\mathcal{S}$, and p' satisfy Axioms I through V and Axiom VI as well. Moreover if we identify each A in $\mathcal{O}$ with Q^A in $\mathcal{O}'$ then $\mathcal{O}$ becomes a part of $\mathcal{O}'$ and when p' is restricted to $\mathcal{O}$ it coincides with p. In other words, Axiom VI may always be made to hold by enlarging $\mathcal{O}$. The physical significance of this addition is easy to understand. If we have a set of questions $E \to q_E$, one for each Borel subset E of the line, which satisfies conditions (a), (b), and (c), we can *define* an observable by identifying the asking of the question q_E with asking: "Did the measurement of the observable give a value in the set E?"

Now let us look at the function $Q \to m_\alpha(Q)$. For each α in $\mathcal{S}$, m_α has the following properties:

(a) If $Q_1, Q_2, \ldots$ are questions such that $Q_i \,\delta\, Q_j$ for $i \neq j$, then $m_\alpha(Q_1 + Q_2 + \cdots) = m_\alpha(Q_1) + m_\alpha(Q_2) + \cdots$.

(b) $m_\alpha(0) = 0$, $m_\alpha(1) = 1$.
(c) $0 \leq m_\alpha(Q) \leq 1$ for all Q.

If m is any real-valued function defined on the set $\mathcal{Q}$ of all questions and satisfies (a), (b), and (c) above, we shall call it a *probability measure on the questions*. The mapping $\alpha \rightarrow m_\alpha$ sets up a one-to-one correspondence between the states and certain probability measures on the questions. It would be tempting to add an axiom stating that every probability measure on the questions is the m_α for some state α. Such a statement is obviously true in the case of classical mechanics. It is also true (but far from obvious) in the model we shall finally adopt for quantum mechanics. If we were to add such an axiom our system would be completely described by the partially ordered set $\mathcal{Q}$ of questions and the operation $Q \rightarrow 1 - Q$. We could then identify $\mathcal{O}$ with the set of all question-valued measures, $\mathcal{S}$ with the set of all probability measures on the questions, and recover p as the function taking q, m, E into $m(q_E)$. Finally even if such an axiom did not hold we could enlarge $\mathcal{S}$ so that it did. Nevertheless we shall resist the temptation. States cannot be constructed physically in the same way that observables can, and it is difficult to think of a plausible physical justification for the axiom.

In any event our system can now be described by giving three things —the partially ordered set $\mathcal{Q}$, the operation $Q \rightarrow 1 - Q$, and a certain subset of the probability measures on $\mathcal{Q}$. To spell this out let us go in the opposite direction and construct a triple $\mathcal{O}$, $\mathcal{S}$, p starting from an abstract, partially ordered set with a "complementation" and certain "probability measures."

Let $\mathcal{L}$ denote an arbitrary, partially ordered set and let $a \rightarrow a'$ denote an involutory anti-automorphism of $\mathcal{L}$; that is, a function from $\mathcal{L}$ to $\mathcal{L}$ such that $a'' = a$ and $a_1 \leq a_2$ implies that $a_2' \leq a_1'$. Let us say that $a_1 \perp a_2$ if $a_1 \leq a_2'$. We call $a \rightarrow a'$ an *orthocomplementation* if it has the following additional properties:

(a) If $a_1, a_2, \ldots$ are members of $\mathcal{L}$ such that $a_i \perp a_j$ for $i \neq j$ then there exists a unique least element a such that $a \geq a_i$ for all i. [We shall denote this element by $a_1 \cup a_2 \cup \cdots$.]
(b) $a \cup a' = b \cup b'$ for all a and b. [We denote $a \cup a'$ by 1.]
(c) If $a \leq b$ then $b = a \cup (b' \cup a)'$.

It is clear that $Q \rightarrow 1 - Q$ is an orthocomplementation in $\mathcal{Q}$.

If $a \rightarrow a'$ is any orthocomplementation in $\mathcal{L}$ then we define a *probability measure* on $\mathcal{L}$ (with respect to $'$) to be a function m from $\mathcal{L}$ to $[0, 1]$ such that $m(1) = 1$, $m(1') = 0$, and $m(a_1 \cup a_2 \cup \cdots) = m(a_1) + m(a_2) + \cdots$ whenever $a_i \perp a_j$ for $i \neq j$.

If $a_1 \leq a_2$ and m is a probability measure then $m(a_2) = m(a_1 \cup (a_2' \cup a_1)') = m(a_1) + m((a_2' \cup a_1)') \geq m(a_1)$. If for some family

$\mathcal{F}$ of probability measures the converse is true, that is, if $m(a_1) \leq m(a_2)$ for all m in $\mathcal{F}$ implies $a_1 \leq a_2$, then $\mathcal{F}$ will be said to be a *full* family. If

$$\sum_{i=1}^{\infty} t_i m_i$$

is in $\mathcal{F}$ whenever $t_i \geq 0$,

$$\sum_{i=1}^{\infty} t_i = 1$$

and $m_i \in \mathcal{F}$, then $\mathcal{F}$ will be said to be ***strongly convex***. Clearly the m_α for α in $\mathcal{S}$ form a strongly convex, full set of probability measures on $\mathcal{Q}$. Finally let us define an *$\mathcal{L}$-valued measure* L on the line to be a function $E \to L_E$ from $\mathcal{B}$ to $\mathcal{L}$ such that

(a) $E \cap F = \phi$ implies $L_E \perp L_F$.

(b) $L_{E_1 \cup E_2 \cup \ldots} = L_{E_1} \cup L_{E_2} \cup \cdots$ whenever $E_i \cap E_j = \phi$ for $i \neq j$.

(c) $L_\phi = 1'$, $L_R = 1$.

The following theorem is then easily proved.

Theorem: Let $a \to a'$ denote an orthocomplementation on an arbitrary partially ordered set $\mathcal{L}$. Let $\mathcal{S}$ denote any full strongly convex family of probability measures on $\mathcal{L}$ (with respect to $'$). Let $\mathcal{O}$ denote the set of all $\mathcal{L}$-valued measures on the real line. For each triple L, m, E, where L is in $\mathcal{O}$, m is in $\mathcal{S}$, and E is in $\mathcal{B}$, let $p(L, m, E) = m(L_E)$. Then $\mathcal{O}, \mathcal{S}, p$ satisfy Axioms I through VI. Moreover there is a one-to-one order preserving map of the $\mathcal{Q}$ for this system onto $\mathcal{L}$ of such a nature that if a and Q correspond then a' and $1 - Q$ do also.

Somewhat more informally stated, the notion of a system $\mathcal{O}, \mathcal{S}, p$ satisfying Axioms I through VI is equivalent to the notion of an orthocomplemented partially ordered set $\mathcal{L}$ together with a full strongly convex family of probability measures on $\mathcal{L}$. The point of the reduction is that the possibilities for systems $\mathcal{L}, \mathcal{S}$ are easier to survey than possibilities for systems $\mathcal{O}, \mathcal{S}, p$. We shall call $\mathcal{Q} = \mathcal{L}$ the *logic*† of our system.

The orthocomplemented partially ordered set $\mathcal{L}$ of all questions plays the role in our system played by phase space in classical mechanics. Then analogy becomes particularly close if we think of classical phase space as the Boolean algebra of all its Borel subsets instead of as a set of points.

---

†Cf. G. Birkhoff and J. von Neumann, "The Logic of Quantum Mechanics," *Annals of Math.*, **37**, 835 (1936).

Thus far our axioms apply both to classical mechanics and to quantum mechanics. We shall ultimately single out quantum mechanics by making an assumption about the structure of $\mathcal{Q}$. First, however, we shall introduce some useful notions which can be discussed without making any special assumptions about this structure.

Let A be an observable and let E be a Borel subset of the real line. If $Q_E^A = 0$, that is, if $\alpha_A(E) = 0$, for all α in $\mathcal{S}$ we shall say that E is of A measure zero or of measure zero with respect to A. Let $\mathcal{I}_A$ denote the set of all open intervals I that are of A measure zero. The union of all of these is an open set $\mathcal{O}_A$. It is an easy exercise in real variable theory to prove that $\mathcal{O}_A$ is itself of A measure zero and contains every *open* set of A measure zero. The *closed* set S_A consisting of all real numbers not in $\mathcal{O}_A$ we shall call the *spectrum* of A. Any point x for which $Q_{\{x\}}^A \neq 0$ is clearly in S_A. The set of all such points we shall call the *point spectrum* of A. If the point spectrum of A is a Borel set whose complement has A measure zero we shall say that A has a *pure point spectrum*. The quantization rules of quantum theory will be seen to arise from the fact that certain key observables have non-empty point spectra.

We shall say that the observable A is *bounded* if S_A is contained in a finite interval. The least positive number N such that $|x| \leq N$ for all x in S_A will be called the *norm* of the observable and is denoted $\|A\|$.

Let A be any bounded observable and let α be any state. Then α_A is a probability measure concentrated in the interval $-\|A\| \leq x \leq \|A\|$. Hence $\int_{-\infty}^{\infty} x \, d\alpha_A(x)$ exists. Following the usual usage of probability theory we shall call the value of the integral the *expected value* or *mean value* of A in the state α. We shall denote it by $m_\alpha(A)$. If A is a question then α_A is concentrated in the points $0, 1$ and $\int_{-\infty}^{\infty} x \, d\alpha_A(x) = 0\alpha_A(\{0\}) + 1\alpha_A(\{1\}) = \alpha_A(\{1\})$. Thus m_α applied to questions agrees with our previous usage of this symbol.

Let us assume for the moment that the following axiom related to Axiom II holds in our system:

Axiom II′: If $m_\alpha(A) = m_\alpha(B)$ for all states α then $A = B$.

Then given two bounded observables A_1 and A_2 there will exist at most one observable A_3 such that

$$m_\alpha(A_3) = m_\alpha(A_1) + m_\alpha(A_2) \qquad \text{for all states } \alpha$$

It is natural to call A_3 the *sum* $A_1 + A_2$ of A_1 and A_2 whenever it exists. Thus in the presence of Axiom II′ we may define what we mean by $A_1 + A_2$ although this sum need not exist. If $A_1 = f_1(B)$ and $A_2 = f_2(B)$ then it is easy to see that $A_1 + A_2$ exists and is equal to $(f_1 + f_2)(B)$. If $A_1 + A_2$ does exist then it is not difficult to show that $\|A_1 + A_2\| \leq \|A_1\| + \|A_2\|$. Moreover it is practically obvious that when the appropriate sums exist we have $(A + B) + C = A + (B + C)$, $A + B = B + A$

with similar identities for multiplication by real numbers. Thus if $A_1 + A_2$ always exists then the bounded observables form a normed vector space.

The assumption that $A + B$ always exists is basic in the formulation of quantum mechanics given by Segal.† His starting point is the normed vector space of all bounded observables. It follows from our axioms that for every state α, $m_\alpha(A^2) \geq 0$ for all observables A, and from the definition of sum that $m_\alpha(\lambda A + \mu B) = \lambda m_\alpha(A) + \mu m_\alpha(B)$. For Segal a state *is* any real-valued function on the bounded observables that has these two properties. From our point of view this definition permits too many states. Not every state in Segal's sense assigns a probability distribution to every bounded observable. On the other hand, Segal's states are presumably all limits of states that do assign probability distributions to all observables, and in applications it is often convenient to deal with such idealized "limit states."

Whenever $A + B$ is always defined we can define "symmetrized products" of bounded observables by setting $A \circ B = [(A+B)^2 - A^2 - B^2]/2$. In the model actually in use for quantum mechanics $A \circ B$ is distributive with respect to addition and the set of all bounded observables forms a commutative, non-associative algebra of a sort known to algebraists as a *Jordan algebra*. Jordan is a physicist who began the study of these algebras in a quantum-mechanical context. Unfortunately there are no known physical reasons for assuming the multiplication $A \circ B$ to be distributive in general, and Lowdenslager and Sherman have given examples showing that distributivity does not follow from Segal's axioms.

Let Q_1 and Q_2 be questions in classical mechanics, i.e., Borel subsets of $\mathcal{M}_{V^*}$. Let $R_1 = Q_1 - Q_1 \cap Q_2$, $R_2 = Q_2 - Q_2 \cap Q_1$, $Q_3 = Q_1 \cap Q_2$. Then R_1, R_2, and Q_3 are mutually disjoint and $Q_1 = R_1 + Q_3$, $Q_2 = R_2 + Q_3$. In the general case this may or may not happen. That is, it may or may not be possible to find three questions R_1, R_2, and Q_3 such that $R_1 \,\delta\, R_2$, $R_2 \,\delta\, Q_3$, $R_2 \,\delta\, Q_3$, $Q_1 = R_1 + Q_3$, $Q_2 = R_2 + Q_3$. Whenever it is possible we shall say that the questions are compatible or *simultaneously answerable*. The justification for this terminology arises from the easily proved fact that Q_1 and Q_2 are simultaneously answerable in the above sense if and only if there exists an observable A and members E_1 and E_2 of $\mathcal{B}$ such that $Q_{E_1}^A = Q_1$ and $Q_{E_2}^A = Q_2$. More generally, we shall define two observables A and B to be simultaneously observable if Q_E^A and Q_F^B are simultaneously answerable for all pairs E and F in $\mathcal{B}$. In our final axiom system it will be possible to prove that two observables A and B are simultaneously observable if and only if there exists an observable C and Borel functions f and

† I. E. Segal, *Annals of Math.*, 48, 930-948 (1947); "A Mathematical Approach to Elementary Particles and Their Fields," lecture notes, University of Chicago, 1955, Chap. 1.

g such that $A = f(C)$ and $B = g(C)$. It is quite possible that this can be proved using the axioms we have already.

Suppose that our system is such that any two questions Q_1 and Q_2 are simultaneously answerable. Then, as is easy to see, our system is a *lattice* in the sense that for each two elements Q_1 and Q_2 there is a least (called $Q_1 \cup Q_2$) that is greater than or equal to each and a greatest (called $Q_1 \cap Q_2$) that is less than or equal to each. Moreover this lattice is a Boolean algebra in the sense that it has an involutory anti-automorphism $Q \to 1 - Q$ such that $Q \cap (1 - Q) = 0$, $Q \cup (1 - Q) = 1$, and satisfies the *distributive law*: $Q_1 \cap (Q_2 \cup Q_3) = (Q_1 \cap Q_2) \cup (Q_1 \cap Q_3)$. Conversely, it is easy to see that if our system is a Boolean algebra then any two questions are simultaneously answerable.

Even if we have non-simultaneously answerable questions there will be at least two questions (0 and 1) that are compatible with every question. Let us call the set of all questions with this property the *center* of $\mathcal{Q}$, and denote it by $C(\mathcal{Q})$. It is easy to see that $C(\mathcal{Q})$ is always a Boolean algebra. Suppose that $C(\mathcal{Q})$ contains a member Q_0 not equal to 0 or 1. Then $1 - Q_0$ is also in $C(\mathcal{Q})$ and every other question R is uniquely of the form $R_1 + R_2$, where $R_1 \leq Q_0$ and $R_2 \leq 1 - Q_0$. Thus $\mathcal{Q}$ is the "direct sum" of the two partially ordered sets $\mathcal{Q}_1$ and $\mathcal{Q}_2$, where $\mathcal{Q}_1$ is the set of all Q with $Q \leq Q_0$ and $\mathcal{Q}_2$ is the set of all Q with $Q \leq 1 - Q_0$. $\mathcal{Q}_1$ and $\mathcal{Q}_2$ satisfy all of our axioms and we may consider their centers and decompose further. In this way we may attempt to write $\mathcal{Q}$ as a direct sum $\mathcal{Q}_1 \oplus \mathcal{Q}_2 \oplus \cdots$, where the $\mathcal{Q}_j$ are *indecomposable* in the sense that their centers are trivial. Of course the example of classical mechanics shows there is no *discrete* decomposition in general. On the other hand, it is almost certain that the modern theory of continuous direct sums could be applied to show that any $\mathcal{Q}$ is a "direct integral" or "continuous direct sum" of indecomposable $\mathcal{Q}$'s. Of course such a continuous sum would be a Boolean algebra if and only if its indecomposable components were trivial, i.e., consisted of one point sets.

From the point of view taken in this course the fundamental difference between quantum mechanics and classical mechanics is that in quantum mechanics there are non-simultaneously answerable questions, i.e., $\mathcal{Q}$ is not a Boolean algebra. One can go much further than this and state not only what $\mathcal{Q}$ is not but what it is. Almost all modern quantum mechanics is based implicitly or explicitly on the following assumption which we shall state as an axiom:

Axiom VII: The partially ordered set of all questions in quantum mechanics is isomorphic to the partially ordered set of all closed subspaces of a separable, infinite dimensional Hilbert space.

This axiom has rather a different character from Axioms I through VI. These all had some degree of physical naturalness and plausibility. Axiom VII seems entirely *ad hoc*. Why do we make it? Can we justify making it? What else might we assume? We shall discuss these questions in turn. The first is the easiest to answer. We make it because it "works," that is, it leads to a theory which explains physical phenomena and successfully predicts the results of experiments. It is conceivable that a quite different assumption would do likewise but this is a possibility that no one seems to have explored.† Ideally one would like to have a list of physically plausible assumptions from which one could deduce Axiom VII. Short of this one would like a list from which one could deduce a set of possibilities for the structure of $\mathcal{Q}$, all but one of which could be shown to be inconsistent with suitably planned experiments. At the moment such lists are not available and we are far from being forced to accept Axiom VII as logically inevitable. On the other hand, this axiom is by no means as completely *ad hoc* as it appears at first sight. If we replace the question, "What must $\mathcal{Q}$ be?" by the easier one, "What might we expect $\mathcal{Q}$ to be, given that natural laws are usually simple and elegant?", we find that we are led to a number of possibilities for Axiom VII that do not differ very much from the one we have adopted.

In this spirit let us investigate ways in which a partially ordered set may be almost a Boolean algebra. A Boolean algebra we recall is a complemented distributive lattice. That our partially ordered set is complemented and, in fact, orthocomplemented follows from our earlier axioms. In the interests of regularity it is reasonable to suppose that it is a lattice. We seek some weakening of the distributive law: $a \cap (b \cup c) = (a \cap b) \cup (a \cap c)$. In the special case in which $a \geq c$ this takes the form $a \cap (b \cup c) = (a \cap b) \cup c$. The converse is *not* true. There exist many non-distributive lattices for which $a \geq c$ implies $a \cap (b \cup c) = (a \cap b) \cup c$. Such lattices are called *modular* and have been extensively studied. The complemented modular lattices may be regarded as the most regularly behaved, partially ordered sets after the Boolean algebras. Certainly they are the most regular that have been seriously studied. Moreover when suitable finiteness conditions are imposed one knows a great deal about the possible complemented modular lattices. Let $\mathfrak{F}$ be any field (not necessarily commutative). Let $V_{\mathfrak{F},n}$ denote the vector space of all n-tuples of elements of $\mathfrak{F}$. Then the partially ordered set $\mathcal{L}_{\mathfrak{F},n}$ of subspaces of $V_{\mathfrak{F},n}$ is a complemented modular lattice and one has the following theorem: Let $\mathcal{L}$ be a complemented modular lattice in which "chains" $a_1 < \cdots < a_n$ have bounded length. Then $\mathcal{L}$ is a direct sum of lattices of the form $\mathcal{L}_{\mathfrak{F},n}$ and of certain lattices in which no chain has length greater than 4. Although no such complete general theorems are available in the

† But see recent work of Jauch, Stueckelberg, and others at the University of Geneva on real and quaternionic Hilbert spaces.

absence of chain conditions,† it is at least strongly suggested that it is reasonable to suppose $\mathcal{Q}$ to be a direct sum or direct integral of suitable, infinite, dimensional generalizations of the $\mathcal{L}_{\mathcal{F},n}$. The most obvious such generalization is the lattice of all closed subspaces of a Banach space. Moreover, it has been proved by the author and S. Kakutani that if the lattice of all closed subspaces of a real or complex Banach space has an orthocomplementation then the norm may be changed to an equivalent one under which the Banach space is a Hilbert space and the orthocomplementation is the obvious one. In short, we are led in a natural way to consider as the most likely candidates for $\mathcal{Q}$ direct sums and direct integrals of the lattices $\mathcal{L}^R$ and $\mathcal{L}^C$, where $\mathcal{L}^R$ is the lattice of all closed subspaces of a real Hilbert space and $\mathcal{L}^C$ is the lattice of all closed subspaces of a complex Hilbert space. If we shut our eyes for the moment to more exotic possibilities for the summands we see that the only serious arbitrariness in our Axiom VII lies in supposing that we have only *one* summand and that this summand is defined by a complex rather than by a real Hilbert space.

The assumption of only one summand amounts physically to the assumption that *every* non-constant observable fails to be simultaneously observable with some observable—a decision that nature might well have made. Moreover on the grounds of mathematical simplicity it is reasonable to begin by considering this case. On the other hand, recent work in "super-selection rules" by Wightman, Wick, and Wigner suggests that at least in some contexts $\mathcal{Q}$ is a direct sum and that there are non-constant universally compatible observables.

It is somewhat harder to see the significance of choosing our Hilbert space to be complex rather than real. However as we shall see in the next section, the existence of a multiplication by i makes it possible to carry over to quantum mechanics one of the most striking formal features of classical mechanics—namely, the natural correspondence between observables and one-parameter groups of symmetries.

Although $\mathcal{L}^R$ and $\mathcal{L}^C$ may be the most obvious generalizations of $\mathcal{L}_{\mathcal{F},n}$, they are by no means the only ones. The reals and complexes are not the only number systems over which one can define Hilbert spaces, and whereas most other possibilities can probably be ruled out on physical grounds for being disconnected or something equally unlikely, at least the quaternions remain and we must add the lattice of closed subspaces of a quaternionic Hilbert space as a possible alternative to $\mathcal{L}^C$. There are also possibilities of a different sort. $\mathcal{L}^C$ can alternatively be defined as the lattice of all self-adjoint idempotents

---

†However, see I. Kaplansky, Any Orthocomplemented Complete Modular Lattice Is a Continuous Geometry, *Annals of Math.*, **61**, 524-541 (1955).

in the ring of all bounded linear operators on a complex Hilbert space. This ring is an example of what Murray and von Neumann call a factor. In finite dimensions all factors are of this kind, but in infinite dimensions only the so-called factors of type I are. For each factor of type II_1, II_∞, and III, one can form the lattice of all self-adjoint idempotents and one must consider it as a possible model for $\mathcal{Q}$. Indeed, von Neumann had just such an application in mind in developing the theory of factors. Moreover when the factor is of type II_1 the lattice in question is actually modular. We should perhaps have mentioned earlier that in spite of our motivating considerations $\mathcal{L}^R$ and $\mathcal{L}^C$ are not quite modular. The strict modular law seems a little too strong unless strong finiteness conditions obtain.

It would be interesting to have a thorough study of the consequences of modifying Axiom VII along the lines indicated by the immediately preceding considerations. However, such a study has not been made and we now end our apology and proceed to deduce the consequences of Axiom VII as stated.

First of all it follows at once from the theorem of Kakutani and the author, mentioned above, that the isomorphism described in Axiom VII may always be chosen so that if Q corresponds to the closed subspace M then $1 - Q$ corresponds to the orthogonal complement $M^\perp$. Moreover, it will be convenient to identify each closed subspace with the projection on that subspace. In other words we may suppose given a one-to-one correspondence between the members of $\mathcal{Q}$ and the projections on a separable, infinite dimensional Hilbert space $\mathcal{H}$ of such a character that if $P \sim Q$ then $1 - P \sim 1 - Q$ and if $Q_1 \sim P_1$ and $Q_2 \sim P_2$ then $Q_1 \leq Q_2$ if and only if $P_1 P_2 = P_2 P_1 = P_1$.

Our next task is to specify which measures on the questions are to be regarded as states and we shall see, thanks to a deep theorem of A. M. Gleason, that there is only one reasonable way to do this. We begin with some examples of measures on the questions. Let ϕ be any unit vector in our Hilbert space $\mathcal{H}$. For each question $\mathcal{Q}$ let $m_\phi(Q) = (P(\phi), \phi)$, where P is the corresponding projection. It is trivial to verify that m_ϕ is a probability measure on $\mathcal{Q}$. Further examples can be obtained by taking convex linear combinations $\gamma_1 m_{\phi_1} + \gamma_2 m_{\phi_2} + \cdots$, where $\gamma_i \geq 0$, $\sum \gamma_i = 1$, and the ϕ_i are unit vectors. What about others? According to Gleason's theorem there are no others. Given any probability measure m on $\mathcal{Q}$ there exists a sequence $\phi_1, \phi_2, \ldots$ of elements of $\mathcal{H}$ of unit length (which can be taken to be orthogonal) and a sequence $\gamma_1, \gamma_2, \ldots$ of positive real numbers such that $\gamma_1 + \gamma_2 + \cdots = 1$ and such that $m = \gamma_1 m_{\phi_1} + \gamma_2 m_{\phi_2} + \cdots$. Gleason's proof of this highly nontrivial theorem will be found in the *Journal of Mathematics and Mechanics*.†

We must decide which of these measures represent states. As re-

†A. M. Gleason, *J. Math. Mechanics*, 6, 885-893 (1953).

marked earlier there is no *a priori* reason for assuming that they all do. However thanks to Gleason's theorem we can *prove* that they all do if we add one new very plausible axiom:

Axiom VIII: If Q is any question different from 0 there exists a state α such that $m_\alpha(Q) = 1$.

Theorem: Every measure on the questions arises from a state.

Proof: Let ϕ be any unit vector in $\mathcal{H}$. Let P be the projection on the one-dimensional subspace generated by ϕ. By Axiom VIII there exists a state α such that $m_\alpha(Q) = 1$, where Q is the question corresponding to P. By Gleason's theorem

$$m_\alpha = \gamma_1 m_{\phi_1} + \gamma_2 m_{\phi_2} + \cdots$$

for some choice of γ_i and ϕ_i. Thus

$$0 = m_\alpha(1 - Q) = \gamma_1(1 - (P(\phi_1), \phi_1)) + \gamma_2(1 - (P(\phi_2), \phi_2)) + \cdots$$

Hence $\gamma_j(1 - (P(\phi_j), \phi_j)) = 0$ for all j. Hence for every j, $P(\phi_j) = \phi_j$. Hence $\phi_j = c_j\phi$ for all j, where $|c_j| = 1$. Hence

$$m_{\phi_j}(P') = (P'\phi_j, \phi_j) = (P' c_j\phi_j, c_j\phi_j) = (P'\phi, \phi)$$

for all P'. Hence $m\phi_j = m_\phi$ for all j. Hence $\sum \gamma_j m_{\phi_j} = m\phi$. Hence $m_\alpha = m_\phi$. Hence m_ϕ arises from a state. Since ϕ was an arbitrary unit vector and every measure on the questions is of the form $\sum \gamma_j m_{\phi_j}$ it follows from Axiom IV that every measure on the questions arises from a state.

It is easy to see that the states that define measures on the questions of the form m_ϕ, where ϕ is a unit vector in H, are all *pure* states. It is obvious that no other states are pure. Thus $\phi \rightarrow m_\phi$ defines a mapping of the unit vectors in Hilbert space onto the pure states of our system. This mapping is not one-to-one, but we verify at once that $m_\phi = m_\psi$ if and only if $\phi = c\psi$ for some complex number c with $|c| = 1$. Thus if we identify ϕ with $e^{ix}\phi$ we have a one-to-one correspondence between unit vectors and pure states.

Now let us see how observables look in the light of Axiom VII. By the general theory each observable defines and is defined by a question-valued measure and all question-valued measures occur. On the other hand, we have now identified the questions with the projections in our Hilbert space. Thus the observables correspond one-to-one to the projection-valued measures in our Hilbert space. But by the spectral theorem there is a natural one-to-one correspondence between projection-valued measures and (not necessarily bounded) self-adjoint operators. Combining these two correspondences we set up a one-to-one correspondence between observables and self-adjoint operators. Let A be any self-adjoint operator in $\mathcal{H}$, let ϕ be any unit vector, and let E be a Borel set. What is the probability that in the pure state defined by ϕ a measurement of the observable defined by A will lead to a value in E? To answer this we first pass to the projection-valued measure P^A associated with A by the spectral theorem. The projection associated with the question "Does the value of the observable lie in E?" is then P_E^A. The probability is then $(P_E^A \phi, \phi)$. Summarizing we now have the fundamental assertion of quantum statics.

The observables correspond one-to-one to the self-adjoint operators in a separable, infinite dimensional Hilbert space $\mathcal{H}$. *The pure states correspond one-to-one to the one-dimensional subspaces of* $\mathcal{H}$. *To find the probability distribution of the observable defined by the self-adjoint operator* A *in the pure state defined by a one-dimensional subspace choose any unit vector* ϕ *in the one-dimensional subspace and let* P^A *be the projection-valued measure associated with* A *by the spectral theorem. Then the desired probability distribution is*

$$E \to (P_E^A(\phi), \phi)$$

Every state is a (possibly infinite) convex combination of pure states.

The introduction of self-adjoint operators may seem a little gratuitous at this point since we pass immediately back to the fundamental projection-valued measures in order to get the statistical properties of an observable. However as we shall now see, the operators themselves play a rather fundamental role in the development of the theory, and in applications it is the operators which are given us and the spectral resolutions which must be found.

As a first example of how the operators may play a direct role let us compute the expected value of the observable defined by the operator A in the state defined by ϕ. This will be $\int_{\infty}^{+\infty} x \, d\alpha(x)$, where α is the relevent probability measure, i.e., $E \to (P_E^A \phi, \phi)$. But

$$\int_{-\infty}^{+\infty} x \, d(P_x^A \phi, \phi) = (A\phi, \phi)$$

by the spectral theorem. Thus the expected value in question is $(A\phi, \phi)$ and may be expressed directly in terms of A and ϕ without the intervention of P^A. Of course if A is unbounded then $A\phi$ need not be defined, but, since an unbounded observable need not have an expected value in every state, we do not expect $(A(\phi), \phi)$ to be defined for all ϕ.

Let A_1 and A_2 be bounded self-adjoint operators and suppose that $(A_1\phi, \phi) = (A_2\phi, \phi)$ for all ϕ. Then replacing ϕ by $\theta + c\psi$ and letting c take on various complex values we deduce easily that $A_1(\theta)\cdot\psi = A_2(\theta)\cdot\psi$ for all θ and ψ and hence that $A_1 = A_2$. Since, as is easily seen, the bounded observables correspond exactly to the bounded self-adjoint operators we deduce that Axiom II′ is satisfied and we may ask whether the sum of two bounded observables exists. From the identity

$$((A + B)(\phi), \phi) = (A(\phi), \phi) + (B(\phi), \phi)$$

we conclude at once that any two bounded observables have a sum and that the operator corresponding to this sum is the sum of the operators to which they correspond. Since two unbounded self-adjoint operators may have nothing but zero in the intersection of their domains, we cannot speak of the sum of two unbounded observables except in special cases.

Let A' be an observable and let A be the self-adjoint operator to which it corresponds. Let f be any Borel function on the line. Then we have defined an observable $f(A')$ in connection with Axiom III and a self-adjoint operator $f(A)$ in connection with our discussion of the spectral theorem. Is $f(A)$ the self-adjoint operator corresponding to the observable $f(A')$? It is easy to see that it is. Indeed the projection-valued measure associated with $f(A)$ is $E \to P^A_{f^{-1}(E)}$ and the question-valued measure associated with $f(A')$ is $E \to Q^{A'}_{f^{-1}(E)}$. Since $Q^{A'}_F$ and P^A_F correspond for all F, $f(A')$ and $f(A)$ must correspond.

Let ϕ be an eigenvector of the self-adjoint operator A and let λ be the corresponding eigenvalue. Then $P^A_{\{\lambda\}}\phi = \phi$, so $(P^A_E\phi, \phi) = 1$ if $\lambda \in E$ and is zero otherwise. In other words, an eigenvector of a self-adjoint operator defines a state in which the corresponding observable takes on the corresponding eigenvalue with probability 1. Conversely, if $(P^A_{\{\lambda\}}\phi, \phi) = 1$ for some ϕ, A, and λ, then $P^A_{\{\lambda\}}(\phi) = \phi$ and it follows easily that $A(\phi) = \lambda\phi$. If A has a pure point spectrum and $\phi_1, \phi_2, \ldots$ is a basis of eigenvectors with corresponding eigenvalues $\lambda_1, \lambda_2, \ldots$ then every unit vector ϕ may be written uniquely in the form $c_1\phi_1 + c_2\phi_2 + \cdots$. We compute easily that in the state defined by ϕ the observable corresponding to A takes on the value λ_j with probability $|c_j|^2$, and it takes on a value in the complement of $\{\lambda_1, \lambda_2, \ldots\}$ with probability 0.

If λ is a point in the continuous spectrum of the operator A then there will be no state in which λ is taken on with a positive probability. On the other hand, given any $\epsilon > 0$, $P^A_{[\lambda - \epsilon, \lambda + \epsilon]}$ will be different from zero and any unit vector in the range of the projection will define a pure state in which the observable has probability 1 of taking on a value between $\lambda - \epsilon$ and $\lambda + \epsilon$. We leave the proof of these facts to the reader.

Let A and B be self-adjoint operators. It is a theorem in operator theory that if A and B are bounded then $AB = BA$ if and only if $P^A_E P^B_F = P^B_F P^A_E$ for all E and F. Since the latter condition makes sense whether or not A and B are bounded we take it as the definition of commuting in the general case. We can then say that A and B commute if and only if the corresponding observables are simultaneously observable in the sense defined earlier. Actually in the present context we can give a somewhat more satisfactory justification of our definition of "simultaneously observable." If A and B commute then it follows from a general theorem that there exists a self-adjoint operator C and Borel functions f and g such that $A = f(C)$ and $B = g(C)$. Thus the observables A' and B' are functions of the same observable C' and are certainly simultaneously observable. Conversely if A' and B' are bounded and are simultaneously observable in a sense which allows us to speak of their product then the operator corresponding to this product must be

$$\frac{(A+B)^2 - A^2 - B^2}{2} = \frac{AB + BA}{2}$$

and that corresponding to $A'^2 B'$ must be both

$$\left[A\frac{(AB+BA)}{2} + \frac{(AB+BA)}{2}A\right]\Big/2 = \frac{A^2B + 2ABA + BA^2}{4}$$

$$\text{and} \quad \frac{A^2B + BA^2}{2}$$

Thus $ABA = A^2B + BA^2/2$ and the same must be true for $f(A)$ and $g(B)$ for arbitrary functions f and g. Thus $P_E^A P_F^B P_E^A = (P_E^A + P_F^B + P_F^B P_E^A)/2$ for all E and F. Multiplying by P_E^A first on the left and then on the right we get two equations that on subtraction yield $P_E^A P_F^B = P_F^B P_E^A$, so that A and B do commute.

When A and B do not commute there are limitations to the degree to which the probability distributions of the corresponding observables may be simultaneously concentrated near to single points. We can get a quantitative measure of the degree of dispersion or "spreadoutness" δ of an observable in a given state by taking the square root of the expected value of the square of the difference between the observable and its expected value. If A is the relevant operator and ϕ the unit vector defining the state, then the dispersion $\delta(A, \phi)$ is such that

$$\begin{aligned}
\delta^2(A,\phi) &= [[A - (A(\phi),\phi)]^2\ \phi,\phi] \\
&= ((A^2 - 2(A(\phi),\phi)A + (A\phi,\phi)^2)\ \phi,\phi) \\
&= (A^2\phi,\phi) - 2(A(\phi),\phi)^2 + (A(\phi),\phi)^2 \\
&= (A^2(\phi),\phi) - (A(\phi),\phi)^2
\end{aligned}$$

Let A and B now be two self-adjoint operators and let ϕ be in both of their domains, and such that $A(\phi)$ is in the domain of B and vice versa. We shall derive a lower bound for the product $\delta(A,\phi)\delta(B,\phi)$. We have

$$2\,[\text{Imaginary part of } (A(\phi), B(\phi))] = (A(\phi), B(\phi)) - (B(\phi), A(\phi))$$

$$= (BA(\phi), \phi) - (AB(\phi), \phi) = ((BA - AB)(\phi), \phi)$$

Hence

$$|(BA - AB)(\phi), \phi| \leq 2|(A(\phi), B(\phi))| \leq 2\|A(\phi) - a\phi\| \|B(\phi) - b\phi\|$$

where a and b are arbitrary real numbers because $(B-b)(A-a) - (A-a)(B-b) = BA - AB$. But

$$\delta^2(A, \phi) = (A^2\phi, \phi) - (A(\phi), \phi)^2 = ((A - (A(\phi), \phi))^2 \phi, \phi)$$

$$= ((A - (A(\phi), \phi))\phi, (A - (A(\phi), \phi)\phi))$$

$$= \|(A - (A(\phi), \phi))\phi\|^2$$

so

$$\delta(A, \phi) = \|(A - (A(\phi), \phi))\phi\|$$

Thus

$$|(BA - AB)(\phi), \phi| \leq 2\,\delta(A, \phi)\delta(B, \phi)$$

and the product of the two dispersions is bounded below by $|((BA - AB)(\phi), \phi)|/2$. As we shall see below, if A is the operator corresponding to a rectangular coordinate and B that corresponding to the time derivative of this coordinate, then $i(AB - BA)$ is a scalar multiple of the identity, the scalar c being inversely proportional to the mass associated with the coordinate. Thus in every state $\delta(A, \phi)\delta(B, \phi) \geq c/2$. For masses of everyday size, $\sqrt{c/2}$ is much smaller than the dispersion caused by ordinary experimental errors. Thus for all practical purposes both observables can have their probability distributions simultaneously concentrated in points. However, for masses of atomic size, c is much larger and one dispersion cannot be made reasonably small without making the other unreasonably large. The relationship $\delta(A, \phi)\delta(B, \phi) \geq c/2$ is a precise form of the famous Heisenberg uncertainty principle. The c turns out to be $h/2\pi m$, where h is Planck's constant.

We conclude this section by giving a more elegant description of mixed states in terms of operators. Let A be a bounded self-adjoint operator with a pure point spectrum. Let $\phi_1, \phi_2, \ldots$ be a complete orthonormal set of eigenvectors and let $A(\phi_j) = \lambda_j\phi_j$. If $|\lambda_1| + |\lambda_2| + \cdots < \infty$ we shall say that A is a trace operator and set Trace $(A) = \lambda_1 + \lambda_2 + .$

More generally let C be any bounded linear operator and let $A = (C + C^*)/2$, $B = (C - C^*)/2i$. Then A and B are bounded and self-adjoint and $C = A + iB$. We say that C is a trace operator if both A and B are and set Trace (C) = Trace (A) + i Trace (B). It is not hard to show that whenever C is a trace operator and $\phi_1, \phi_2, \ldots$ is a basis for $\mathcal{H}$ then Trace C = $(C(\phi_1), \phi_1) + (C(\phi_2), \phi_2) + \cdots$ where the series converges absolutely. It follows immediately that Trace $(C_1 + C_2)$ = Trace (C_1) + Trace (C_2) whenever all three traces exist. It is also not difficult to prove that $C_1 + C_2$, SC_1, and C_1S are all trace operators whenever C_1 and C_2 are and S is any bounded linear operator. Moreover Trace (SC_1) = Trace (C_1S).

Now let ϕ be a unit vector in $\mathcal{H}$ and let P_ϕ denote the operator $\psi \to (\psi, \phi)\phi$. We see at once that P_ϕ is the projection on the one-dimensional space of all multiples of ϕ. Hence it is a trace operator of trace 1. In general a projection on a *finite* dimensional subspace will be a trace operator whose trace is the dimension of this subspace. If S is any bounded linear operator then $(P_\phi S(\phi), \phi) = (S(\phi), \phi)(\phi, \phi) = (S(\phi), \phi)$ and for ϕ' orthogonal to ϕ, $(P_\phi S(\phi'), \phi') = 0$. Hence taking ϕ to be part of a complete orthonormal set we find that Trace $(P_\phi S)$ = $(S(\phi), \phi)$. Hence for every projection P, $m_\phi(P)$ = Trace $(P_\phi P)$ = Trace (PP_ϕ). More generally, let A be any bounded self-adjoint trace operator whose spectrum is non-negative and whose trace is 1. Let $\phi_1, \phi_2, \ldots$ be its eigenvectors. Then $A(\phi_j) = \gamma_j \phi_j$, where $\gamma_j \geq 0$ and $\gamma_1 + \gamma_2 + \cdots = 1$. Then Trace (PA) = Trace (AP) = $\gamma_1 m_{\phi_1}(P) + \gamma_2 m_{\phi_2}(P) + \cdots$. Thus P → Trace (AP) defines a measure on the questions, and hence a state, for every A with the properties listed. The converse is also true. If α is any state then $m_\alpha = \gamma_1 m_{\phi_1} + \gamma_2 m_{\phi_2} + \cdots$, where $\gamma_i \geq 0$ and $\gamma_1 + \gamma_2 + \cdots = 1$ but the ϕ_j need not be orthogonal. However $A = \gamma_1 P_{\phi_1} + \gamma_2 P_{\phi_2} + \cdots$ is a non-negative self-adjoint operator of trace 1 and Trace (AP) = $m_\alpha(P)$ for all projections P. To summarize: The states may be put into one-to-one correspondence with the non-negative self-adjoint trace operators of trace 1 in such a manner that if α and A correspond then $m_\alpha(P)$ = Trace (PA) for every projection P.

We note that the expected value of the bounded observable defined by the self-adjoint operator B in the mixed state $\gamma_1 \alpha_{\phi_1} + \cdots$ is $\gamma_1(B(\phi_1), \phi_1) + \cdots = \gamma_1$ Trace $(BP_{\phi_1}) + \cdots$ = Trace $(B(\gamma_1 P_{\phi_1} + \cdots))$ = Trace (BA), where A is the operator defining the state.

It is interesting that in describing states by operators we have a one-to-one correspondence, in contradistinction to what took place when we represented pure states by unit vectors. To see what happened note that when $|c| = 1$ the *operators* P_ϕ and $P_{c\phi}$ are identical. Our ambiguity arises when we attempt to describe a projection on a one-dimensional subspace by a particular unit vector in the subspace.

The mixed states and the operator description of them were introduced into quantum mechanics by von Neumann. They play a central role in quantum statistical mechanics. The matrix of an operator defining a mixed state (with respect to an appropriate basis) is known to physicists as a von Neumann density matrix.

2-3 Quantum Dynamics and the Schrödinger Equation

Up to this point we have been discussing quantum statics— the relationship between states and observables at a particular instant of time. In this section we shall discuss quantum dynamics—the way in which states change with the passage of time. In spite of the acausal character of quantum mechanics, as reflected in the impossibility of making more than statistical statements about the results to be expected from the measurements of observables, the relationship between states at different times is strictly causal. That is, the state at time $t_2 > t_1$ is uniquely determined by $t_2 - t_1$ and the state at time t_1. Just as in classical mechanics we have a one-parameter semi-group $t \to V_t$ of transformations of $\mathcal{S}$ into $\mathcal{S}$, such that if the state of our system is α at time t_1 it is $V_{t_2 - t_1}(\alpha)$ at time $t_2 > t_1$. Let $\alpha_1, \alpha_2, \ldots$ be states and let $\gamma_1, \gamma_2, \ldots$ be positive real numbers whose sum is 1. Suppose that at $t = 0$ we are in the state α_j with probability γ_j. Then at $t = t_0$ we are in the state $V_{t_0}\alpha_j$ with probability γ_j. Hence if we are in the state $\gamma_1\alpha_1 + \cdots$ at $t = 0$ we are in the state $\gamma_1 V_{t_0}(\alpha_1) + \cdots$ at time $t = t_0$. Thus $V_t(\gamma_1\alpha_1 + \cdots) = \gamma_1 V_t(\alpha_1) + \cdots$. In other words each V_t must preserve convex combinations of states.

These physical considerations combined with the natural expectation that for each fixed triple A, α, E in $\mathcal{O} \times \mathcal{S} \times \mathcal{B}$ the probability $p(A, \alpha, E)$ should change only slightly in a short time interval, leads to the following addition to our axiom system. We are given a one-parameter semi-group $t \to V_t$ of transformations of $\mathcal{S}$ into $\mathcal{S}$ such that we have:

Axiom IX: For each sequence $\alpha_1, \alpha_2, \ldots$ of members of $\mathcal{S}$ and each sequence $\gamma_1, \gamma_2, \ldots$ of non-negative real numbers whose sum is 1, $V_t(\gamma_1\alpha_1 + \cdots) = \gamma_1 V_t(\alpha_1) + \cdots$ for all $t \geq 0$ and for all A in $\mathcal{O}$, α in $\mathcal{S}$, and E in $\mathcal{B}$ $p(A, V_t(\alpha), E)$ is a continuous function of t.

As in classical mechanics we shall say that our system is *reversible* if each V_t is invertible. Defining $V_{-t} = V_t^{-1}$, $t \to V_t$ becomes a one-parameter group. We shall deal only with the reversible case.

In view of the latter part of the last section we may identify $\mathcal{S}$ with the set of all self-adjoint operators with trace 1 and non-negative spectrum. Thus each V_t becomes a mapping of operators on operators. Let U be any unitary operator in $\mathcal{H}$. It is obvious that $A \to UAU^{-1}$ maps $\mathcal{S}$ onto $\mathcal{S}$ in such a way as to preserve convex combinations. The same is true if U is an *anti-unitary operator*, i.e., a conjugate linear $(U(\lambda\phi) = \bar{\lambda}U(\phi))$ inner product preserving mapping of $\mathcal{H}$ on $\mathcal{H}$. On the other hand, using results of

Kadison† it can be shown that every convex combination preserving one-to-one map of $\mathcal{S}$ on $\mathcal{S}$ is of the form $A \to UAU^{-1}$, where U is either unitary or anti-unitary. Moreover if U_1 and U_2 define the same transformation of $\mathcal{S}$ into $\mathcal{S}$ then $U_1^{-1}U_2$ commutes with all A in $\mathcal{S}$. It is very easy to show from this that $U_1^{-1}U_2$ is a constant multiple of the identity. Thus $U_1 = cU_2$ where $|c| = 1$, since U_1 and U_2 are both unitary.

For each t choose U_t so that $V_t(A) = U_tAU_t^{-1}$. Then

$$V_{t_1+t_2}(A) = U_{t_1+t_2}A\,U_{t_1+t_2}^{-1} = V_{t_1}V_{t_2}(A) = U_{t_1}U_{t_2}AU_{t_2}^{-1}U_{t_1}^{-1}$$

Hence $U_{t_1+t_2} = c(t_1, t_2)U_{t_1}U_{t_2}$, where $|c(t_1, t_2)| = 1$. In particular $U_{2t} = c(t, t)\,U_t^2$, so $U_t = c(t/2, t/2)\,U_{t/2}^2$. Since the square of an anti-unitary transformation is unitary it follows that U_t is unitary for all t. If we apply U_t to the pure state P_ϕ, where ϕ is a unit vector, then $U_tP_\phi U_t^{-1}(U_t(\phi)) = U_t(\phi)$ so that $U_tP_\phi U_t^{-1} = P_{U_t(\phi)}$. Thus the pure state represented by the unit vector ϕ changes in time t to the pure state represented by the unit vector $U_t(\phi)$. Now $p(A, \alpha_\phi, E) = \alpha_\phi(P_E^A) = (P_E^A(\phi), \phi)$. Hence $p(A, U_t(\alpha_\phi), E) = (P_E^AU_t(\phi), U_t(\phi))$. Hence for every projection P, $(P\,U_t(\phi), U_t(\phi))$ is continuous in t for all ϕ. Hence for all ϕ and ψ $(P_\psi U_t(\phi), U_t(\phi))$ is continuous in t by Axiom IV. On the other hand, $(P_\psi U_t(\phi), U_t(\phi)) = (U_t(\phi), \psi)(\psi, U_t(\phi)) = |(U_t(\phi), \psi)|^2$. Hence $|(U_t(\phi), \psi)|^2$ is continuous in t for all ϕ. Without changing V_t we may replace each U_t by $a(t)\,U_t$ where $|a(t)| = 1$. It is possible to show that this choice may be made so that $(U_t(\phi), \psi)$ is continuous in t and so that $c(t_1, t_2) = 1$. (We omit the rather lengthy detailed argument.) In other words, we may find a *continuous one-parameter unitary group* $t \to U_t$ such that $V_t(A) = U_tA\,U_t^{-1}$ for all t and A. If a is any real number then $t \to e^{-iat}U_t$ defines the same V but this is the extent of the ambiguity. U_t is uniquely determined up to multiplication by e^{-iat}. We shall refer to $t \to U_t$ as the *dynamical group of our system*. Its existence may be regarded as the fundamental statement of *general* quantum dynamics.

Applying Stone's theorem to the unitary group $t \to U_t$ we may write it in the form $t \to e^{-iHt}$, where H is a self-adjoint operator and $-iH$ is the infinitesimal generator of $t \to U_t$. The self-adjoint operator H, which completely determines the dynamics of our system, we shall call the *dynamical operator*. It plays the same role in quantum mechanics that the Hamiltonian function plays in classical mechanics. Like the latter it is arbitrary up to an additive constant. Of course H is uniquely determined by U_t but changing U_t to $e^{-iat}U_t$ has the effect of adding a times the identity to H.

Since H is a self-adjoint operator it corresponds to some observable

† *Annals of Math.*, 54, 325 (1951).

and this observable is clearly one of central importance. As we shall see, it is a constant multiple of the quantum mechanical analog of the total energy.

Let ϕ_0 be in the domain of H and be a unit vector. Then at time t the pure state represented by the vector ϕ_0 at $t = 0$ is represented by $U_t(\phi_0) = e^{-iHt}\phi_0$. Let $\phi_t = e^{-iHt}\phi_0$. Then ϕ_t as a function of t satisfies the differential equation

$$\frac{d}{dt}\phi_t = -iH(\phi_t)$$

This is the abstract form of the celebrated Schrödinger equation. It plays the role in quantum mechanics played by Hamilton's equation in classical mechanics. It is a first-order differential equation (in an infinite dimensional space) whose solutions are the trajectories of the pure states. Later we shall throw it into concrete form by realizing the vectors ϕ as complex-valued functions of 3n variables and H as a differential operator. Schrödinger's equation will then become a partial differential equation which is of the first order in the time.

Let ϕ be any eigenvector of the operator H. Then $e^{-itH}(\phi) = e^{-t\lambda}\phi$ where $H(\phi) = \lambda\phi$. Thus for all t $e^{-itH}(\phi)$ and ϕ define the same state. In other words, the state defined by ϕ is a *stationary state* in the sense that it is fixed in time. Conversely if $e^{-itH}(\phi) = \rho(t)\phi$ so that ϕ defines a pure state we see easily that $\rho(t) = e^{-it\lambda}$ so that $H(\phi) = \lambda\phi$. The stationary pure states are precisely the pure states defined by the eigenvectors of the dynamical operator H.

Of course H need not have any eigenvectors at all. Consider the contrary case in which it has a pure point spectrum. Let $\phi_1, \phi_2, \ldots$ be a complete orthonormal set of eigenvectors. Let $H(\phi_j) = \lambda_j\phi_j$. If ψ represents any pure state we have

$$\psi = c_1\phi_1 + c_2\phi_2 + \cdots$$

and

$$U_t(\psi) = c_1 e^{-i\lambda_1 t}\phi_1 + c_2 e^{-i\lambda_2 t}\phi_2 + \cdots$$

In this way we have a very explicit solution of Schrödinger's equation.

Let us call a quantum mechanical observable A' an *integral* if its probability distribution in every state is constant in time. That is, if $p(A', \alpha_{U_t(\phi)}, E)$ is independent of t for all ϕ and E. This means that $(P_E^A U_t(\phi), U_t(\phi)) = (U_t^{-1} P_E^A U_t(\phi), \phi)$ is independent of t and hence

$$(U_t^{-1} P_E^A U_t(\phi), \phi) = (P_E^A \phi, \phi)$$

for all t, E, and ϕ. Hence U_t commutes with P_E^A for all t and E.

Now, by general theorems in operator theory U_t commutes with P_E^A for all E and t if and only if A and H commute. Thus the integrals are just the observables whose corresponding operators commute with H.

Now H itself is a self-adjoint operator and as such corresponds to an observable. Since H commutes with itself this observable is an integral. This integral obviously plays an important role in the theory. We shall see later that it is a constant multiple of the quantum analog of the energy integral of classical mechanics. Thus the stationary pure states of a quantum-mechanical system are just those in which the energy has a definite value with probability 1.

Given an observable that is not an integral we can compute the rate of change of its expected value in various states. Let A be the self-adjoint operator corresponding to the observable and let ϕ be a unit vector in the domain of A. Then the expected value at time t of the observable defined by ϕ at $t = t_0$ is $(AU_t\phi, U_t\phi) = (U_t^{-1} A U_t\phi, \phi)$. A formal computation valid for suitably restricted ϕ leads to the value $i((HA - AH)\phi, \phi)$ for the derivative of this value at $t = 0$. When A and H are both bounded operators these calculations are universally valid and we see that there is a unique observable, namely, that defined by the self-adjoint operator $i(HA - AH)$, which has the property that in every state its expected value is the time derivative of the expected value of the observable defined by A in that state. It is natural to call this new observable the derivative of the old with respect to time. When A and H are unbounded, difficulties arise owing to the fact that H and A may not have domains that overlap suitably. Whenever they do, that is, whenever there is a *unique* self-adjoint operator that agrees with $i(HA - AH)$ for all ϕ for which $HA - AH$ is defined we shall say that the observable defined by A is differentiable with respect to time and that its derivative is the observable corresponding to the unique self-adjoint extension of $i(HA - AH)$.

The notion of time derivative for an observable takes on a somewhat more transparent form if we change our point of view somewhat. Let A be a self-adjoint operator, let ϕ be a unit vector, and let W be a unitary transformation. Then the probability distribution $E \to (P_E^A W(\phi), W(\phi))$ of the observable defined by A in the state defined by $W(\phi)$ is identical with the probability distribution $(W^{-1} P_E^A W(\phi), \phi)$ of the observable defined by $W^{-1}AW$ in the state defined by ϕ. In other words, if we think of the states as being fixed and the observables as changing in time according to the law $A \to U_t^{-1} A U_t$ then we will get the same answers as before for the changes of probabilities with time. The physicists distinguish between these two view points by referring, respectively, to the Schrödinger picture (variable states) and the Heisenberg picture (variable observables). In the Heisenberg picture the notion of the time derivative of an observable is an immediate and obvious one.

Now all that we have done for the one-parameter group $t \to V_t$ applies equally well to any other continuous one-parameter group $t \to W_t$ of "automorphisms" of $\mathcal{S}$. Each W_t is of the form $A \to U_t^W A(U_t^W)^{-1}$, where $t \to U_t^W$ is a continuous one-parameter unitary group uniquely determined by W up to multiplication by a complex-valued function of the form e^{-iat}. By Stone's theorem U_t^W is uniquely determined by its infinitesimal generator $-iA^W$. Thus if we identify two observables whenever they differ by a constant we have a one-to-one correspondence between observables, on the one hand, and continuous one-parameter groups of automorphisms, on the other. This is analogous to the correspondence in classical mechanics between certain differentiable observables, on the one hand, and one-parameter groups of contact transformations, on the other. In that case also two observables that differed by a constant defined the *same* one-parameter group of contact transformations. Just as with U_t, we may define derivatives of observables with respect to U_t^W and get the formula $i(AB - BA)$ for the operator corresponding to the derivative of the observable corresponding to B whenever $-iA$ is the infinitesimal generator of U^W. Thus the expressions $i(AB - BA)$ play the role in quantum mechanics played by the Poisson brackets in classical mechanics. In particular the observable corresponding to A is an integral if and only if H is invariant under the group of automorphisms defined by A.

2-4 The Canonical "Quantization" of a Classical System

So far observables have been entirely abstract entities. We turn now to the problem of correlating them with the observables of classical mechanics and of finding the quantum-mechanical analogs or refinements of specific classical systems.

Suppose that our classical system is of the sort discussed in Sec. 1-2, consisting of n "particles" having coordinates $q_1, \ldots, q_{3n}$ in some rectangular coordinate system. The analysis that causes one to reject the possibility of assigning exact simultaneous values to q_i and $\dot{q}_i$ does not apply to the problem of measuring q_i and q_j simultaneously. We suppose, therefore, that for each j there is a self-adjoint operator Q_j in the Hilbert space of the quantum analog of our classical system and that these Q_j commute with one another. We note that this assumption is consistent with the analogy between Poisson brackets and commutators since the Poisson bracket of any two coordinates (indeed of any two functions of the coordinates alone) is identically zero.

Now, clearly, changing the coordinate system that is used in assigning numbers to positions in space will effect a permutation of the observables. For example a shift of the origin two units in the direction of decreasing x will take Q_j into $Q_j + 2I$ for $j = 1, 4, 7, \ldots$, where I is the identity operator. On the other hand, such a shift or any other one-to-one distance preserving of space onto itself will take questions into questions and preserve the partial ordering in $\mathcal{Q}$ as well as the operation $Q \to 1 - Q$. But it is a well-known consequence of the fundamental theorem of projective geometry that any automorphism of the orthocomplemented lattice of all closed subspaces of Hilbert space may be induced by a unitary transformation U which is unique up to a multiplicative constant: $P \to UPU^{-1}$, where P is the projection on the closed subspace in question. It follows in particular that for any translation in space $x, y, z \to x - a,\ y - b,\ z - c$ there exists a unitary transformation $U_{a,b,c}$ such that $U_{a,b,c} Q_j U_{a,b,c}^{-1} = Q_j + aI,\ \ Q_j + bI$, or $Q_j + cI$ according as q_j is an x coordinate, a y coordinate, or a z coordinate.

Actually we can if we like use different coordinate systems in making observations on different particles and change them independently. Thus given any translation $q_j \to q_j - a_j$ in the space $\mathfrak{M}$ of all $3n$-tuples $q_1, \ldots, q_{3n}$ there will be a unitary operator $U_{a_1,\ldots,a_{3n}}$ such that for all j

$$U_{a_1,\ldots,a_{3n}} Q_j U^{-1}_{a_1,\ldots,a_{3n}} = Q_j + a_j I$$

Let us forget for the moment the existence of the $U_{a_1,\ldots,a_{3n}}$ and consider the consequences of the fact that all Q_j commute. Let $E_1 \times \cdots \times E_{3n}$ denote a "rectangular" Borel subset of $\mathfrak{M}$ where E_j is a Borel subset of the line and let $P_{E_1 \times \cdots \times E_{3n}}$ be defined as $P^{Q_1}_{E_1} \cdots P^{Q_{3n}}_{E_{3n}}$. Since the $P^{Q_j}_{E_j}$ commute with one another, $P_{E_1 \times \cdots \times E_{3n}}$ is a projection. Moreover it is not hard to show that the mapping $E_1 \times \cdots \times E_{3n} \to P_{E_1 \times \cdots \times E_{3n}}$ has a unique extension to a projection-valued measure $E \to P_E$ whose domain is the set of all Borel subsets of $\mathfrak{M}$. [The definition of projection-valued measure on the line has an obvious and immediate extension whose formulation we leave to the reader.] Using this it is easy to assign a meaning to $f(Q_1, \ldots, Q_{3n})$, where f is any real-valued Borel function defined on $\mathfrak{M}$. It is the self-adjoint operator whose projection-valued measure is $E \to P_{f^{-1}(E)}$. Clearly $f(Q_1, \ldots, Q_n)$ commutes with all Q_j. If the converse is true we shall say that the Q_j form a *complete commuting set of operators*. It is easy to see that this is the case if and only if the range of P is a maximal Boolean algebra of projections in the sense that no projection not in this range can commute with everything in it. Moreover it can be shown that a Boolean algebra of projections which is closed under the countable Boolean operations is maximal if and only if there exists a vector ϕ in the Hilbert space such that the finite linear combinations $c_1 P_{E_1}(\phi) + c_2 P_{E_2}(\phi) + \cdots$ are dense in the Hilbert space. ϕ is called a "cyclic vector" for the family.

We now make the tentative assumption that the Q_j do form a complete commuting set of operators. This assumption is suggested by the fact that in classical mechanics any observable having zero Poisson bracket with every coordinate is itself a function of the coordinates. It will have to be modified when we apply quantum mechanics to the atom because of the fact that electrons have a degree of freedom (spin) with no classical counterpart. However, it is convenient to make it now, as it simplifies our work and leads us in the right direction.

Let ϕ be a cyclic vector of the Boolean algebra of all P_E. Then $E \to (P_E(\phi), \phi)$ is a measure μ on $\mathfrak{M}$. Let $E_1, E_2, \ldots, E_s$ be disjoint Borel subsets of $\mathfrak{M}$ and let ψ_{E_j} be the characteristic function of E_j. Then

$$c_1\psi_{E_1} + c_2\psi_{E_2} + \cdots + c_s\psi_{E_s} \to c_1 P_{E_1}(\phi) + \cdots + c_s P_{E_s}(\phi)$$

defines a linear map of a dense subspace of $\mathcal{L}^2(\mathfrak{M}, \mu)$ onto a dense subspace of $\mathcal{H}$. Moreover,

$$\begin{aligned} \|c_1 P_{E_1}(\phi) + \cdots + c_s P_{E_s}(\phi)\|^2 &= \sum c_i \overline{c_j}(P_{E_i}(\phi), P_{E_j}(\phi)) \\ &= \sum c_i \overline{c_j}(P_{E_i \cap E_j}(\phi), \phi) \\ &= \sum |c_i|^2 (P_{E_i}(\phi), \phi) = \sum |c_i|^2 \mu(E_i) \\ &= \int_{\mathfrak{M}} |c_1\psi_{E_1} + \cdots + c_s\psi_{E_s}|^2 \, d\mu \end{aligned}$$

Thus our linear map is norm-preserving. Extending it by continuity we get a unitary map of $\mathcal{L}^2(\mathfrak{M}, \mu)$ on $\mathcal{H}$ and it is obvious that under this map the operator P_E in $\mathcal{H}$ corresponds to the operator $f \to \psi_E f$ in $\mathcal{L}^2(\mathfrak{M}, \mu)$. It follows at once that the operator Q_j in $\mathcal{H}$ corresponds to the operator $f \to q_j f$ in $\mathcal{L}^2(\mathfrak{M}, \mu)$. In other words, assuming that the operators Q_j corresponding to the coordinate observables form a *complete* commuting family, we conclude that $\mathcal{H}$ can be realized concretely as $\mathcal{L}^2(\mathfrak{M}, \mu)$, where μ is some finite measure in the space $\mathfrak{M}$ of all 3n-tuples $q_1, \ldots, q_{3n}$ in such a manner that Q_j becomes the operation of multiplying by q_j.

The measure μ is, of course, not uniquely determined. Indeed, if ν is any measure (finite or only σ – finite) defined on Borel subsets of $\mathfrak{M}$ and having the same null sets as μ, then the mapping

$f \to \sqrt{\rho}\, f$, where ρ is the Radon-Nikodym derivative[†] of μ with respect to ν, will be a unitary map of $\mathcal{L}^2(\mathfrak{M}, \mu)$ on $\mathcal{L}^2(\mathfrak{M}, \nu)$ which commutes with multiplication by any function on $\mathfrak{M}$. On the other hand, this is the extent of the ambiguity. The null sets of μ are precisely the sets E such that $P_E = 0$ and as such are uniquely determined.

Now let us exploit the existence of the $U_{a_1, \ldots, a_{3n}}$. We deduce at once that $U_{a_1, \ldots, a_{3n}} P_E U^{-1}_{a_1, \ldots, a_{3n}} = P_{E'}$, where E' is the result of translating E by $-a_1, \ldots, -a_{3n}$. Hence if $P_E = 0$ then also $P_{E'} = 0$. Hence the null sets of our measure μ must be translation invariant. But it is known than any Borel measure μ on a finite dimensional vector space whose null sets are invariant has the same null sets as Lebesgue measure. Thus we may take μ to be Lebesgue measure.

Before proceeding to identify further observables we remark that there is no reason to suppose that every classical observable has a quantum counterpart or that if it has one it has only one. The general observable in classical mechanics is obtained by taking a general function of the 6n basic observables $q_1, \ldots, q_{3n}, \dot{q}_1, \ldots, \dot{q}_{3n}$, and in quantum mechanics we can only take more or less arbitrary functions of observables whose operators commute. We can get a wide variety of quantum observables by taking sums and applying Borel functions of one variable in succession but they will not correspond in a one-to-one fashion to classical ones. Moreover as classical mechanics is a limiting case of quantum mechanics it would not be surprising to have a number of different quantum observables coincide in the classical limit. On the other hand, we shall find natural correspondences between the *basic* classical observables and corresponding quantum ones.

We suppose our system is such that each Q_j defines an observable differentiable with respect to time in the sense discussed in Sec. 2-3 and let $\dot{Q}_j$ denote the differentiated observable. Formally, $\dot{Q}_j = i(HQ_j - Q_jH)$, where H is the, as yet unknown, dynamical operator of our system. For obvious reasons we call $\dot{Q}_j$ the velocity observable corresponding to the coordinate q_j. Now, in classical mechanics the momentum observables were defined in a manner having no obvious connection with velocities and turned out to be constant multiples of them. The momentum p_j, for example, was the observable associated with the one-parameter group of contact transformations gen-

---

[†]Let μ and ν be two measures in the same space S. The Radon-Nikodym theorem states that whenever certain mild regularity conditions are satisfied and $\mu(E) = 0$ implies $\nu(E) = 0$ there exists a measurable function ρ which is the "density" of ν with respect to μ; that is, $\nu(E) = \int_E \rho(s)\, d\mu(s)$. This function is often called the Radon-Nikodym derivative of ν with respect to μ.

erated by the one-parameter group

$$q_1, \ldots, q_{3n} \to q_1, \ldots, q_{j-1}, q_j - a, q_{j+1}, \ldots, q_{3n}$$

in configuration space. But this group also induces a group of unitary transformations W_a^j in $\mathcal{H} = \mathcal{L}^2(\mathfrak{M}, \mu)$, namely,

$$W_a^j \psi(q_1, \ldots, q_{3n}) = \psi(q_1, \ldots, q_{j-1}, q_j - a, q_{j+1}, \ldots, q_{3n})$$

and this one-parameter unitary group has a skew-adjoint infinitesimal generator $-\partial/\partial q_j$ which corresponds to the self-adjoint operator $1/i\ \partial/\partial q_j$. Let us make the assumption that quantum mechanics follows classical mechanics to the extent that the observable associated with W^j; i.e., the observable whose operator is i times the infinitesimal generator of W^j is a constant c_j times $\dot{Q}_j$. Moreover let us proceed formally, ignoring questions about domains of operators. Then our assumption becomes

$$\dot{Q}_j = i(HQ_j - Q_jH) = 1/ic_j \frac{\partial}{\partial q_j}$$

so

$$HQ_j - Q_jH = -1/c_j \frac{\partial}{\partial q_j}$$

Now

$$\frac{\partial^2}{\partial q_j^2} Q_k = Q_k \frac{\partial^2}{\partial q_j^2} \qquad \text{if } k \neq j$$

and

$$\frac{\partial^2}{\partial q_j^2} Q_j - Q_j \frac{\partial^2}{\partial q_j^2} = \frac{2\partial}{\partial q_j}$$

Hence if we set

$$H_T = -\frac{1}{2} \sum_{j=1}^{3n} 1/c_j \frac{\partial^2}{\partial q_j^2}$$

we compute that $H_TQ_j - Q_jH_T = HQ_j - Q_jH$ for all j. Hence $H - H_T$ commutes with all Q_j. Hence $H - H_T$ is a function of Q_j. Hence $H - H_T$ is multiplication by some real-valued function Y of $q_1, \ldots, q_{3n}$. Hence H has the following form:

$$H(\psi) = -\frac{1}{2}\sum_{j=1}^{3n}\frac{1}{c_j}\frac{\partial^2\psi}{\partial q_j^2} + Y\psi$$

Conversely, of course, if H has this form then $\dot{Q}_j$ will be $1/ic_j\ \partial/\partial q_j$. Now suppose that Y is differentiable. Then $\ddot{Q}_j = i(H\dot{Q}_j - \dot{Q}_jH) =$ multiplication by $-1/c_j\ \partial Y/\partial q_j$ (since $\dot{Q}_j$ commutes with H_T). This means that in any "suitably differentiable" state ψ we have

$$\frac{d^2}{dt^2}(Q_j(\psi),\psi) = -\frac{1}{c_j}\left(\frac{\partial Y}{\partial q_j}\psi,\psi\right)$$

Now suppose that ψ is such that $\int_E|\psi(q_1,\ldots,q_{3n})|^2\,d\mu$ is very near to 1 for some very small set E and that this condition persists for some time (with varying E of course). Then ψ defines a state in which the coordinates $q_1,\ldots,q_{3n}$ have definite values to a high degree of approximation. These definite values may be taken to be $(Q_1\psi,\psi),\ldots,$ $(Q_{3n}\psi,\psi)$. Moreover,

$$\left(\frac{\partial Y}{\partial q_j}\psi,\psi\right) = \int_E\frac{\partial Y}{\partial q_j}|\psi|^2\,d\mu$$

is equal (approximately) to the value of $\partial Y/\partial q_j$ at the point $(Q_1\psi,\psi),$ $\ldots,(Q_{3n}\psi,\psi)$. In other words, our quantum system will approximate a classical one with differential equations

$$c_j\frac{d^2q_j}{dt^2} = -\frac{\partial Y}{\partial q_j}$$

i.e., a classical system with masses $c_1,\ldots,c_{3n}$ and potential energy Y. However our equations can equally well be written in the form

$$c_j\hbar\frac{d^2q_j}{dt^2} = -\hbar\frac{\partial Y}{\partial q_j}$$

where $\hbar$ is any constant. Thus our system also behaves like a classical one with masses $c_1\hbar,\ldots,c_{3n}\hbar$ and potential energy $\hbar Y$. It follows that we cannot at once identify the c_j with the masses but only with some multiple of the masses $c_j = m_j/\hbar$. Similarly, we cannot identify the function Y with the potential energy V but only with the same multiple of V: $Y = V/\hbar$. For any $\hbar$ the quantum system whose dynamical operator is formally such that

$$H(\psi) = -\hbar/2\sum_{j=1}^{3n}1/m_j\frac{\partial^2}{\partial q_j^2} + \frac{V}{\hbar}\psi$$

acts under suitable circumstances like the classical system whose masses are $m_1, \ldots, m_{3n}$ and whose potential energy function is V. The corresponding Schrödinger equation

$$H(\psi) = -1/i \frac{\partial \psi}{\partial t} = -\hbar/2 \sum_{j=1}^{3n} 1/m_j \frac{\partial^2}{\partial q_j^2} + \frac{V}{\hbar}\psi$$

depends essentially on the value of $\hbar$. Even if V is zero so that we have non-interacting particles the $\hbar$ does *not* cancel out. Thus the quantum system has features depending on the value assigned to $\hbar$ which are obliterated upon passage to the classical limit. One of these is the rate at which the probability distributions spread out with the passage of time. The smaller we take $\hbar$ the more slowly they spread. This fact is not immediately obvious but may be deduced from a study of the equations. By making experiments which reveal the non-classical feature of the system we can determine the ratios $m_j/\hbar$. Thus if we know m_j then we can determine $\hbar$. It turns out to be a universal constant closely related to that of Planck. We shall discuss the exact relationship later.

Note that by choosing the unit of mass suitably we can make $\hbar = 1$. If we had not already chosen a unit of mass we could *define* the mass associated with the j-th coordinate to be the number c_j. This is consistent with the usual definition of mass and does not involve an arbitrary choice of a unit of mass. In other words, absolute mass has significance in quantum mechanics and (after choosing units of time and length) we have a "natural" mass unit. The constant $\hbar$ may be looked upon as a conversion factor from the natural mass to the one adopted before the advent of quantum mechanics.

Let us summarize the position we have now reached. Given a classical system of n particles with rectangular coordinates $q_1, \ldots, q_{3n}$, masses $m_1, \ldots, m_{3n}$, and potential function V we may obtain a quantum-mechanical refinement of it as follows. Let $\mathcal{H}$ be the Hilbert space of all square summable complex-valued functions on Euclidean 3n-dimensional space using Lebesgue measure. Let the observable q_j be associated with the multiplication operator $f \to q_j f = Q_j(f)$. Let H be a self-adjoint operator which takes the twice-differentiable function ψ into

$$-\hbar/2 \sum_{j=1}^{3n} 1/m_j \frac{\partial^2 \psi}{\partial q_j^2} + \frac{V}{\hbar}\psi$$

where $\hbar$ is a certain universal constant determined by experiment. Then the evolution of our system in time is given by the one-parameter group e^{-itH}, or, in differential form, our state function ψ varies in time so as to satisfy the Schrödinger equation

$$-1/i = \frac{\partial \psi}{\partial t} = -\hbar/2 \sum_{j=1}^{3n} 1/m_j \frac{\partial^2 \psi}{\partial q_j^2} + \frac{V}{\hbar}\psi$$

Some cautionary remarks need to be made. First of all, we are not assured that we have found the only quantum system having the given classical system as a limit but only that we have found one such system. We shall see that there are others. Moreover, although we have sketched an argument designed to show that it is the only one having certain simple properties, we have not shown that this sketch can be filled out to a rigorous proof.

Second, our description of the quantum system is incomplete in that it is possible that the formal differential operator we have written down may have more than one self-adjoint extension—if it has any at all. On the other hand, there is a convenient way of making the extension in most cases of physical interest and in many of these even a unique one. Moreover, the answers to most physical questions seem to be independent of which extension one takes. Thus the ambiguity is not serious. A detailed discussion in one important special case will be found in a paper by Kato.†

The velocity observable $\dot{Q}_j$ is formally

$$i(HQ_j - Q_jH) = -i\hbar/m_j \frac{\partial}{\partial q_j} = \hbar/im_j \frac{\partial}{\partial q_j}$$

There is a "natural" self-adjoint extension of this: $\hbar/im_j$ times the infinitesimal generator of the one-parameter group U_t where $U_t(\psi(q_1, \ldots, q_{3n})) = \psi(q_1, \ldots, q_{j-1}, q_j - t, q_{j+1}, \ldots, q_{3n})$. Therefore we take $\hbar/im_j$ times this infinitesimal generator as the rigorously defined analog of the velocity observable. With the dynamical operator as well as the operator corresponding to the position and velocity observables all given we have a mathematical model capable in principle of supplying the answer to any physical question we might ask.

Let us now see to what extent we can find quantum analogs for familiar dynamical concepts. We have already indicated why the self-adjoint operator $1/i\ \partial/\partial q_j$ should be regarded as the quantum analog of the classical mechanical momentum observable p_j. On the other hand, since the velocity observable corresponds to the operator $\hbar/im_j\ \partial/\partial q_j$ we should have

$$m_j(\hbar/im_j \frac{\partial}{\partial q_j}) = \hbar/i \frac{\partial}{\partial q_j}$$

for momentum if we are to have numerical agreement between the con-

† T. Kato, *Trans. Am. Math. Soc.*, 70, 195 (1951).

cepts in the classical limit. The discrepency is, of course, due to the fact that momentum in classical mechanics is not defined until a unit of mass is chosen. This is so even if we use the group theoretical definition, since the mapping of $\mathfrak{M}_V$ on $\mathfrak{M}_{V^*}$ depends on the choice of a unit of mass. The discrepency disappears if we use the natural mass unit so that $\hbar = 1$. If we insist upon using other mass units we must define momentum by $\hbar/i\ \partial/\partial q_j$ instead of the more natural $1/i\ \partial/\partial q_j$.

It is not hard to show that the sums

$$\hbar/i \sum_{i=1}^{n} \frac{\partial}{\partial x_i}, \quad \hbar/i \sum_{i=1}^{n} \frac{\partial}{\partial y_i}$$

$$\hbar/i \sum_{i=1}^{n} \frac{\partial}{\partial z_i}$$

where $q_1, \ldots, q_{3n} = x_1, y_1, z_1, \ldots, x_n, y_n, z_n$ define integrals whenever V is such that momentum is conserved in the classical system. Thus we have a momentum conservation law in quantum mechanics as well as in classical mechanics. Indeed it can be derived directly from the same symmetry considerations.

Similar remarks apply to angular momentum. We define the z component of the total angular momentum about the origin to be the observable corresponding to $\hbar/i$ times the infinitesimal generator of the one-parameter group in our Hilbert space generated by the rotations about the z axis. The formal operator turns out to be

$$\hbar/i \sum_{j=1}^{n} \left(x_j \frac{\partial}{\partial y_j} - y_j \frac{\partial}{\partial x_j} \right)$$

and there are similar definitions and expressions for the angular momentum about other axes. Using polar coordinates it is easy to compute directly that the angular momentum operators have pure point spectra, the eigenvalues being integral multiples of $\hbar$. Thus quantum mechanics provides a rationale for one of the fundamental principles of the old quantum theory—provided we take $\hbar$ as $h/2\pi$.

Expressed in terms of rectangular coordinates q_i and the corresponding momenta p_i the total energy in classical mechanics is

$$\frac{1}{2} \sum_{j=1}^{3n} p_j^2/m_j + V(q_1, \ldots, q_{3n})$$

This expression still makes sense if we replace the q's and the p's by the operators corresponding to them in quantum mechanics. $V(Q_1, \ldots, Q_{3n})$ is simply the multiplication operator $\psi \to V(\psi)$ and we are led to the (formal) operator

$$\psi \to -\sum_{j=1}^{3n} \hbar^2/2m_j \frac{\partial^2 \psi}{\partial q_j} + V\psi$$

as the operator corresponding to a possible quantum analog of energy in classical mechanics. We note that it is just $\hbar$ times the formal expression for the dynamical operator. We thus *define* energy in quantum mechanics as the observable corresponding to $\hbar$ times the dynamical operator and note that it is an integral so that we have an energy conservation law. We also have the following convenient rule of thumb for writing down the formal expression for the dynamical operator and hence Schrödinger's equation: (1) Express the classical energy in terms of *rectangular* coordinates and momenta. (2) Replace each p_j by the operator $\hbar/i\ \partial/\partial q_j$ and each q_j by the operator $\psi \to q_j\psi$. (3) Divide by $\hbar$. As we shall see below there is a more fundamental procedure that applies to generalized coordinates.

If we denote the self-adjoint operator corresponding to p_j by P_j and use the fact that P_j is formally $\hbar/i\ \partial/\partial q_j$, we have the following relationships (valid when the unbounded operators Q_j and P_j are suitably restricted):

$$Q_jQ_k - Q_kQ_j = P_jP_k - P_kP_j = 0$$

$$Q_jP_k - P_kQ_j = i\hbar\ \delta_j^k$$

They are known as the Heisenberg commutation relations. They are interesting in that (except for the occurrence of $\hbar$) they are suggested *a priori* by the parallel between Poisson brackets and commutator brackets and that, to an extent that we shall make precise below, they uniquely determine the concrete realizations

$$Q_j(\psi) = q_j\psi \qquad P_j(\psi) = \hbar/i \frac{\partial\psi}{\partial q_j}$$

We have, of course,

$$[q_j, q_k] = [p_j, p_k] = 0 \qquad [q_j, p_k] = \delta_j^k$$

Concerning the uniqueness just alluded to, it is easy to state a precise theorem once the Heisenberg relations have been thrown into the so-called "Weyl form" so that only bounded operators appear. Let

U^j denote the one-parameter unitary group $s \to e^{-isQ_j}$ and let V^j denote the group $t \to e^{-itP_j}$. Then the U^j and V^k determine the Q_j and P_k, and on a formal level the Heisenberg commutation relations are equivalent to the following:

$$
\begin{aligned}
&U_s^j U_t^k = U_t^k U_s^j \qquad V_s^j V_t^k = V_t^k V_s^j \\
&U_s^j V_t^k = V_t^k U_s^j e^{i\delta_j^k \hbar st}
\end{aligned}
\tag{3}
$$

Now let $U^1, \ldots, U^{3n}, V^1, \ldots, V^{3n}$ be *any* 6n continuous one-parameter unitary groups acting in the same separable Hilbert space $\mathcal{H}$ and satisfying (3). According to a theorem of Stone and von Neumann,† $\mathcal{H}$ can be written as a direct sum $\mathcal{H} = \mathcal{H}_1 \oplus \ldots$, where each $\mathcal{H}_j$ is carried into itself by all U_s^j and all V_t^k and where each $\mathcal{H}_i$ can be mapped unitarily on $\mathcal{L}^2(E^{3n})$ in such a manner that the operator U_s^j goes over into $\psi \to e^{-isq_j}\psi$ and the operator V_t^k goes over into

$$\psi(q_1, \ldots, q_{3n}) \to \psi(q_1, \ldots, q_{k-1}, q_k - \hbar t, q_{k+1}, \ldots, q_{3n})$$

Adding the assumption that there is only one summand is completely equivalent to the assumption that the Q_j form a complete commuting family.

When $\mathcal{H}$ has been realized as $\mathcal{L}^2(\mathfrak{M})$ so that the Q_j and P_j are as indicated above, each state vector ψ is a square summable function on $\mathfrak{M}$ and each trajectory $e^{-iHt}(\psi)$ is a function on $\mathfrak{M} \times R$, where R is the real line. This function of $3n + 1$ real variables is called a *wave function* for the system. We may write ψ in the form $\psi = \sqrt{\rho}\, e^{iS}$, where $\rho = |\psi|^2$ and S is a real-valued function on $\mathfrak{M} \times R$ which is determined only up to an additive constant. The functions ρ and S together uniquely determine ψ, and ρ has an obvious physical significance. If E is any Borel subset of $\mathfrak{M}$ then $\int \cdots \int_E (q_1, \ldots, q_{3n})\, dq_1 \ldots dq_{3n}$ is the probability in the state ψ that measurements of the q_j will give a value for the 3n-tuple $q_1, \ldots, q_{3n}$ lying in E. S also has a simple physical meaning which is not quite so obvious. Assuming that ψ is suitably differentiable, let us compute the expected value of the j-th momentum component, i.e., $(P_j(\psi), \psi)$. It is

$$\frac{\hbar}{i} \int \left(\frac{\partial \psi}{\partial q_j} \overline{\psi} \right) dq_1 \ldots dq_{3n}$$

---

†See *Annals of Math.*, **33**, 567 (1932), for the von Neumann proof.

$$= \frac{\hbar}{i} \int \cdots \int \left(\frac{\partial}{\partial q_j} \left(\sqrt{\rho}\, e^{iS} \right) \left(\sqrt{\rho}\, e^{-iS} \right) \right) dq_1 \ldots dq_{3n}$$

$$= \frac{\hbar}{i} \int \cdots \int \left(\sqrt{\rho}\, \frac{\partial \sqrt{\rho}}{\partial q_j} \right) dq_1 \ldots dq_{3n} + \frac{\hbar}{i} \int \rho i \frac{\partial S}{\partial q_j}\, dq_1 \ldots dq_{3n}$$

$$= \frac{\hbar}{i} \int \cdots \int \frac{1}{2} \frac{\partial \rho}{\partial q_j}\, dq_1 \ldots dq_{3n} + \frac{\hbar}{i} \int \cdots \int \rho i \frac{\partial S}{\partial q_j}\, dq_1 \ldots dq_{3n}$$

Since $\rho \to 0$ as $q_j \to \pm\infty$, therefore $\int \cdots \int (\partial\rho/\partial q_j)\, dq_1 \ldots dq_{3n} = 0$, and we see that the expected value of P_j is just $\hbar \int \cdots \int \rho\, (\partial S/\partial q_j)\, dq_1 \ldots dq_{3n}$. If we have a state in which ρ is highly concentrated, i.e., in which $q_1, \ldots, q_{3n}$ is almost sure to be very near to $q_1^0, \ldots, q_{3n}^0$, then the momentum components will have expected values very near to $\partial S/\partial q_j\, (q_1^0, \ldots, q_{3n}^0)$. In any case,

$$q_1, \ldots, q_{3n} \to \hbar \frac{\partial S}{\partial q_1}(q_1, \ldots, q_{3n}) \quad \hbar \frac{\partial S}{\partial q_2}(q_1, \ldots, q_{3n}), \ldots$$

gives a vector field in configuration space that associates a set of momentum values to every set of coordinate values. The mean of these momentum values with respect to ρ is the set of expected values of the momenta in the state $\sqrt{\rho}\, e^{iS}$. In this sense $\hbar[\partial S/\partial q_1, \ldots, \partial S/\partial q_{3n}]$ describes the momentum of the state.

The reader may find it interesting to write Schrödinger's equation as a system of two real equations in S and ρ, especially in the case of one particle. He will get equations that are very closely analogous to the equations of hydrodynamics, with ρ as fluid density and S as velocity potential.

2–5 Some Elementary Examples and the Original Discoveries of Schrödinger and Heisenberg

Consider the case in which $n = 1$ and $V = 0$, the so-called free particle. Let $q_1 = x$, $q_2 = y$, $q_3 = z$ so that

$$H(\psi) = -\frac{\hbar}{2m} \left(\frac{\partial^2 \psi}{\partial x^2} + \frac{\partial^2 \psi}{\partial y^2} + \frac{\partial^2 \psi}{\partial z^2} \right)$$

We can get an explicit spectral analysis of H by using Fourier transforms. Each unit vector in $\mathcal{H} = \mathcal{L}^2(E^3)$ is uniquely of the form

$$\frac{1}{(2\pi)^{3/2}} \iiint_{-\infty}^{\infty} \hat{\psi}(a, b, c)\, e^{i(ax + by + cz)}\, da\, db\, dc$$

where $\hat{\psi}$ is also of $\mathcal{L}^2$-norm 1, and for such a ψ

$$\frac{1}{i}\frac{\partial\psi}{\partial x} = \frac{1}{(2\pi)^{3/2}} \iiint_{-\infty}^{\infty} a\widehat{\psi}(a, b, c) e^{i(ax+by+cz)} da\, db\, dc$$

with similar expressions for $(1/i)(\partial\psi/\partial y)$ and $(1/i)(\partial\psi/\partial z)$. Thus $H(\psi)$ is the Fourier transform of $\hbar/2m(a^2 + b^2 + c^2)\widehat{\psi}(a,b,c)$ and $e^{-iHt}\psi$ is the Fourier transform of $\exp\{-[\hbar i/2m(a^2 + b^2 + c^2)t]\}\widehat{\psi}(a,b,c)$. It follows that, if ψ at time $t = 0$ is the Fourier transform of $\widehat{\psi}(a,b,c)$, then ψ at time t has the form

$$\psi(x, y, z, t) = \frac{1}{(2\pi)^{3/2}} \iiint_{-\infty}^{\infty} \widehat{\psi}(a, b, c) \exp[i(ax+by+cz)]$$

$$\exp\left[-\frac{i\hbar}{2m}(a^2 + b^2 + c^2)t\right] da\, db\, dc$$

Thus $\psi(x, y, z, t)$ is a continuous "superposition" of functions of x, y, z, t of the form

$$\exp\left\{i\left[ax+by+cz - \frac{\hbar}{2m}(a^2+b^2+c^2)t\right]\right\}$$

$$= \exp\left[i\sqrt{a^2+b^2+c^2}\left(\frac{ax+by+cz}{\sqrt{a^2+b^2+c^2}} - \frac{\hbar}{2m}\sqrt{a^2+b^2+c^2}\, t\right)\right]$$

Such a function describes a "plane wave" traveling with velocity $\hbar/2m\sqrt{a^2+b^2+c^2}$ in a direction perpendicular to the plane $ax+by+cz=0$, and having wavelength $2\pi/\sqrt{a^2+b^2+c^2}$. Suppose that $\widehat{\psi}$ is zero outside of a small region about a_0, b_0, c_0. Then the $x, y,$ and z components of momentum are certain to have values very near to $\hbar a_0, \hbar b_0,$ and $\hbar c_0$, respectively, and ψ as a function of x, y, z and t is a superposition of plane waves with wavelength very near to $2\pi/\sqrt{a_0^2+b_0^2+c_0^2} = 2\pi\hbar/p_0$, where $p_0 = \sqrt{(\hbar a_0)^2+(\hbar b_0)^2+(\hbar c_0)^2}$ is a value which the total momentum is certain to be near. Thus in a state where the momentum is certain to be near p_0 the state function is a superposition of plane waves with wavelengths near $2\pi\hbar/p_0$. In this sense a free particle of momentum p is associated with a wave of length $2\pi\hbar/p$.

In 1924, in the last days of the old quantum theory, de Broglie had advanced a speculative theory according to which a particle of momentum p was associated with a "wave" of length h/p. This theory was confirmed experimentally shortly thereafter by various scientists who observed diffraction patterns from beams of electrons of momentum p of the sort one got from x rays of wavelength h/p. Quantum mechanics provides an explanation if $\hbar$ is taken to be $h/2\pi$, and we have a second experimental verification of the equality $\hbar = h/2\pi$.

Inspired by de Broglie's work, E. Schrödinger attempted to found a "wave mechanics" that would be related to classical mechanics as wave optics is related to ray optics. Experimenting with possible wave equations and their associated standing waves and applying his ideas to the electron in the hydrogen atom he was led to consider the equation

$$\frac{\partial^2\psi}{\partial x^2} + \frac{\partial^2\psi}{\partial y^2} + \frac{\partial^2\psi}{\partial z^2} + \frac{2m}{K^2}\left(E + \frac{e^2}{\sqrt{x^2+y^2+z^2}}\right)\psi = 0$$

where E is the energy of the electron, m its mass, e its charge, and K a certain constant. Concerning this equation he made the startling and fundamental discovery that (a) it has solutions with certain reasonable regularity properties only when E had as its value one of the discrete series of values $me^4/2K^2$, $me^4/8K^2$, $me^4/32K^2, \ldots$ and (b) these values corresponded to the Bohr energy levels of the hydrogen atom when K was given the value $h/2\pi$. Thus Schrödinger had found a way of getting these discrete values without making any *a priori* discreteness assumptions. In 1926 and 1927 he published a series of papers entitled (in English translation) "Quantization as an Eigenvalue Problem" in which he exploited and discussed some of the implications of this important discovery.

It is, of course, easy to understand Schrödinger's discovery now. If we take the Rutherford model of the hydrogen atom as a heavy, positively charged nucleus with an electron of equal and opposite charge $-e$ moving in an orbit around it and disregard the motion of the nucleus, we get a classical mechanical system whose energy is

$$\frac{1}{2m}(p_x^2 + p_y^2 + p_z^2) - \frac{e^2}{\sqrt{x^2+y^2+z^2}}$$

Passing to the corresponding quantum system we have the Hilbert space $\mathcal{L}^2(E^3)$ and an energy observable corresponding to the differential operator

$$\psi \to -\frac{\hbar^2}{2m}\left(\frac{\partial^2\psi}{\partial x^2} + \frac{\partial^2\psi}{\partial y^2} + \frac{\partial^2\psi}{\partial z^2}\right) - \frac{e^2\psi}{\sqrt{x^2+y^2+z^2}}$$

The stationary pure states of our system will be those defined by the eigenfunctions of this operator and the corresponding eigenvalues will be the certain values of the energy in these states. But a differentiable function ψ will be an eigenfunction of this operator with eigenvalue E if and only if it satisfies the equation

$$-\frac{\hbar^2}{2m}\left(\frac{\partial^2\psi}{\partial x^2} + \frac{\partial^2\psi}{\partial y^2} + \frac{\partial^2\psi}{\partial z^2}\right) - \frac{e^2\psi}{\sqrt{x^2+y^2+z^2}} = E\psi$$

This is equivalent to

$$\frac{\partial^2\psi}{\partial x^2} + \frac{\partial^2\psi}{\partial y^2} + \frac{\partial^2\psi}{\partial z^2} + \frac{2m}{\hbar^2}\left(E + \frac{e^2}{\sqrt{x^2 + y^2 + z^2}}\right) = 0$$

which is just Schrödinger's equation with $\hbar = K$. The fact that the eigenvalues check with the observed spectrum of hydrogen when one takes $K = h/2\pi$ provides a third confirmation of the identity of $\hbar$ with $h/2\pi$.

A few months earlier a superficially rather different approach to a natural explanation of the energy levels in atoms had been begun by Heisenberg and developed by Heisenberg, Born, and Jordan. Heisenberg wished to develop a new mechanics in which such non-directly observable entities as electron positions and velocities would be eliminated. By vague and mystical but inspired heuristic reasoning he was led to consider analogs of the differential equations of mechanics in which the varying elements were infinite matrices. The element in the n-th row and m-th column of such a matrix was somehow supposed to be associated with a transition from the m-th to the n-th energy level of the system. In the developed theory a matrix Q_i is assigned to each classical coordinate and a matrix P_i to the corresponding momentum. The matrices are required to satisfy the following conditions:

(a) $P_iQ_j - Q_jP_i = (h/2\pi i)\delta^i_j$.

(b) $P_iP_j - P_jP_i = Q_iQ_j - Q_jQ_i = 0$ for all i and j.

(c) The matrix H obtained by substituting the P_i and the Q_j into the classical expression for the energy has zeros for its non-diagonal entries. The diagonal values of H are then taken as the energy levels of the system.

It was found that this matrix problem could be solved for various interesting classical systems and led to the same energy levels as Schrödinger's wave mechanics. To one familiar with the connection between matrices and operators it is at least formally clear why this should be so. The operators that one substitutes into the classical expression for energy to get the Schrödinger operator clearly satisfy conditions (a) and (b). Thus if one introduces a basis with respect to which the Schrödinger operator has a diagonal matrix, the matrices of our operators will satisfy conditions (a), (b), and (c), and the diagonal values will be the eigenvalues of the Schrödinger operator. Moreover as we saw in the last section, the equations (a) and (b) have an essentially unique solution.

These ideas introduced by Heisenberg, Schrödinger, Born, and Jordan attracted widespread attention and many physicists and mathematicians, notably Dirac, Bohr, and von Neumann, contributed to their development and interpretation. By 1929 the efforts of these men had converted the old quantum theory and the crude ideas just described into the systematic refinement of classical mechanics, now known as quantum mechanics.

2-6 Generalized Coordinates

In this section we shall show how the prescription given in Sec. 2-4 for "quantizing" a classical system can be reformulated so as to apply to the more general systems considered in Sec. 1-3. We begin with some purely mathematical considerations.

Let $\mathfrak{M}$ be a set in which there is given a notion of "Borel set," i.e., in which there is given a family of subsets closed under complementation and countable unions, which we shall call the Borel sets. Let C be a measure class in $\mathfrak{M}$, that is, the set of all σ-finite measures having the same null sets as one of them. For each α in C we may form the Hilbert space $\mathcal{L}^2(\mathfrak{M}, \alpha)$. Let $T_{\alpha_1\alpha_2}$ denote the mapping $f \to \sqrt{d\alpha_1/d\alpha_2}\, f$. Clearly $T_{\alpha_1\alpha_2}$ is a unitary map of $\mathcal{L}^2(\mathfrak{M}, \alpha_1)$ onto $\mathcal{L}^2(\mathfrak{M}, \alpha_2)$. Moreover since $T_{\alpha_2\alpha_3}T_{\alpha_1\alpha_2} = T_{\alpha_1\alpha_3}$ we see that we have a mutually consistent family of canonical mappings of the $\mathcal{L}^2(\mathfrak{M}, \alpha)$ upon one another. It follows that we may think of the $\mathcal{L}^2(\mathfrak{M}, \alpha)$ as being one Hilbert space dependent upon C alone and not upon the choice of one of its members—a member of this one Hilbert space being a family of mutually corresponding members of the $\mathcal{L}^2(\mathfrak{M}, \alpha)$, one from each.

This Hilbert space may be described in a much less cumbersome way as follows. Given f in $\mathcal{L}^2(\mathfrak{M}, \alpha)$ form the finite measure $|f|^2\alpha$ and the function $f/|f|$. In this way we map $\mathcal{L}^2(\mathfrak{M}, \alpha)$ *onto* the set of all pairs α_1, h, where α_1 is a finite Borel measure whose null sets include those of C, and h is a Borel function from $\mathfrak{M}$ to the unit circle $|z| = 1$. Of course, we must identify α_1, h with α_1, h' whenever $h = h'$ almost everywhere with respect to α_1. Now the mapping $f \to |f|^2\alpha, f/|f|$ is one-to-one. Moreover, if $g = T_{\alpha_1\alpha_2}(f)$, where f is in $\mathcal{L}^2(\mathfrak{M}, \alpha_1)$ and g is in $\mathcal{L}^2(\mathfrak{M}, \alpha_2)$, then $|f|^2\alpha_1, f/|f| = |g|^2\alpha_2, g/|g|$ since $g = \sqrt{d\alpha_1/d\alpha_2}\, f$. In other words the pair $|f|^2\alpha, f/|f|$ is independent of α as f varies from one $\mathcal{L}^2(\mathfrak{M}, \alpha)$ to another and may be taken as the general member of an "intrinsic" Hilbert space $\mathcal{H}_C$. We compute easily that the inner product of $|f_1|^2\alpha, f_1/|f_1|$ with $|f_2|^2\alpha, f_2/|f_2|$, which by definition must be $\int_{\mathfrak{M}} f_1\bar{f}_2\, d\alpha$, may be thrown into the form

$$\int_{\mathfrak{M}} (f_1/|f_1|)(\bar{f}_2/|f_2|)\, |f_1|\, |f_2|\, d\alpha$$

$$= \int_{\mathfrak{M}} (f_1/|f_1|)(\bar{f}_2/|f_2|) \sqrt{(d|f_1|^2\alpha/d\alpha)(d|f_2|^2\alpha/d\alpha)}\, d\alpha$$

These motivating remarks having been made, we proceed to give a straightforward formal definition of the intrinsic Hilbert space $\mathcal{H}_C$ of a measure class C. Let α_1 and α_2 be any two finite measures defined on all Borel subsets of $\mathfrak{M}$. It is trivial that the measure $E \to \int_E \sqrt{(d\alpha_1/d\alpha_3)(d\alpha_2/d\alpha_3)}\, d\alpha_3$ is independent of α_3, where α_3 is any finite measure defined on all Borel subsets of $\mathfrak{M}$ with respect to which both α_1 and α_2 are absolutely continuous. We shall denote this measure by $\sqrt{\alpha_1\alpha_2}$. Now let $\mathcal{H}_C$ be the set of all pairs α, h, where α is a finite measure defined on all Borel subsets of $\mathfrak{M}$ and absolutely continuous with respect to the members of C, h is a Borel function from $\mathfrak{M}$ to the unit circle $|z| = 1$, and α, h is identified with α, h' whenever $h = h'$ almost everywhere with respect to α. It follows immediately from the considerations of the preceding paragraphs that there is one and only one way of converting $\mathcal{H}_C$ into a Hilbert space in such a manner that the inner product of α_1, h_1 with α_2, h_2 is $\int h_1 \bar{h}_2 d\sqrt{\alpha_1 \alpha_2}$. Moreover $f \to |f|^2 \alpha, f/|f|$ is a unitary map of $\mathfrak{L}^2(\mathfrak{M}, \alpha)$ on $\mathcal{H}_C$ for all α in C.

Now let $\mathfrak{M}$ be a C_∞ manifold and let the Borel sets be those defined by the topology. If we introduce a local coordinate system $q_1, \ldots, q_n$ in the open set $\mathcal{O}$ we get a measure in $\mathcal{O}$ by using the mapping $x \to q_1(x), \ldots, q_n(x)$ to transfer Lebesgue measure from E^n. The *null sets* of this measure are independent of the coordinate system used and it is not hard to show that there exist measures in $\mathfrak{M}$ whose null sets in each $\mathcal{O}$ are just those defined by local coordinate systems. We shall call the measure class of these measures the *natural measure class* of $\mathfrak{M}$. The Hilbert space of the natural measure class we shall call the *intrinsic Hilbert space of the manifold* and denote by $\mathcal{H}_{\mathfrak{M}}$.

Let us call a measure α in $\mathfrak{M}$, a C_∞ measure, if in any local coordinate system its Radon-Nikodym derivative with respect to the measure defined by the coordinate system is a C_∞ function. The set of all α, f in $\mathcal{H}_{\mathfrak{M}}$ such that α and f are both C_∞ can be shown to be a dense subspace of $\mathcal{H}_{\mathfrak{M}}$ which we denote by $\mathcal{H}^0_{\mathfrak{M}}$ and call the C_∞ subspace of $\mathcal{H}_{\mathfrak{M}}$.

If V is any automorphism of $\mathfrak{M}$ as a C_∞ manifold, then V defines a one-to-one mapping of $\mathcal{H}_{\mathfrak{M}}$ on $\mathcal{H}_{\mathfrak{M}}$ which is obviously unitary (and maps $\mathcal{H}^0_{\mathfrak{M}}$ on $\mathcal{H}^0_{\mathfrak{M}}$). In particular every C_∞ one-parameter group $t \to V_t$ of automorphisms of $\mathfrak{M}$ defines a one-parameter unitary group $t \to U_t$ on $\mathcal{H}_{\mathfrak{M}}$. By Stone's theorem we may write $U_t = e^{Kt}$, where K is skew-adjoint. We wish to relate the skew-adjoint operator K in $\mathcal{H}_{\mathfrak{M}}$ to the contravariant vector field L which is the infinitesimal generator of V_t. To do this let α be a C_∞ member of the natural measure class of $\mathfrak{M}$ and realize $\mathcal{H}_{\mathfrak{M}}$ as $\mathfrak{L}^2(\mathfrak{M}, \alpha)$. A straightforward computation shows that for all f in C_∞ with compact support we have $K(f) = L(f) - (\sigma/2)f$, where σ is the derivative at $t = 0$ of the square root of the Radon-Nikodym derivative with respect to α of the transform of α by V_t. In particular when α is invariant under the V_t we have $L(f) = K(f)$. The function σ can be shown to be the unique real-valued function such that the operator $f \to L(f) - (\sigma/2)f$ is formally skew-symmetric in $\mathfrak{L}^2(\mathfrak{M}, \alpha)$, i.e., such that

$$\int L(f)g\, d\alpha + \int L(g)f\, d\alpha + \int \sigma fg\, d\alpha = 0$$

As such it can be defined whether or not L is the infinitesimal generator of a one-parameter group. When $q_1, \ldots, q_n$ is a local coordinate system such that $\alpha(E) = \int_E dq_1 \ldots dq_n$ and $L = \Sigma\, a_i(\partial/\partial q_i)$ then

σ turns out to be $\Sigma\, \partial a_i/\partial q_i$. Thus it is consistent with the usage of classical vector analysis to call σ the *divergence* of L with respect to α. We shall write $\sigma = \mathrm{div}_\alpha L$.

Now let T be a C_∞ Riemannian metric in $\mathfrak{M}$. For each C_∞, ψ on $\mathfrak{M}$ $d\psi$ will be a C_∞ covariant vector field in $\mathfrak{M}$ and T will convert it into a C_∞ contravariant vector field $\widetilde{d\psi}^T$. On the other hand, there is a C_∞ measure α_T in $\mathfrak{M}$ canonically associated with T. We shall not attempt to describe it precisely (though this would not be hard) but content ourselves with the remark that it is the one which assigns n-dimensional "volumes" consistent with the one-dimensional "lengths" defined by integrating $T(d\phi/dt,\ d\phi/dt)$ along the curves $t \to \phi(t)$. The operator $\psi \to \mathrm{div}_{\alpha_T}(\widetilde{d\psi}^T)$ is then a second-order differential operator canonically associated with T. When $\mathfrak{M}$ is Euclidean space, T is the usual Euclidean metric, and $q_1, \ldots, q_n$ is a rectangular coordinate system this operator reduces to $\psi \to (\partial^2\psi/\partial q_1^2) + \cdots + (\partial^2\psi/\partial q_n^2)$, i.e., to the familiar Laplacian. In general it is called the *Laplace-Beltrami operator* of T. Applied to C_∞ functions with compact support it is a symmetric operator with respect to the inner product in $\mathcal{L}^2(\mathfrak{M}, \alpha_T)$. Hence via the canonical mapping of $\mathcal{L}^2(\mathfrak{M}, \alpha_T)$ on $\mathcal{H}_{\mathfrak{M}}$ it defines a symmetric operator in $\mathcal{H}_{\mathfrak{M}}$. We shall denote this operator by ∇^T.

If f is any real-valued Borel function on $\mathfrak{M}$, let Q_f denote the self-adjoint operator on $\mathcal{H}_{\mathfrak{M}}$ which takes α_1, h into $|f|^2\alpha_1, h$ and is defined whenever $\int |f|^2\, d\alpha_1 < \infty$. Of course if we realize $\mathcal{H}_{\mathfrak{M}}$ as $\mathcal{L}^2(\mathfrak{M}, \alpha)$ then Q_f is simply multiplication by f.

Now let $\mathfrak{M}$ be E^{3n} and let H be the Hamiltonian function defined by a potential V and masses $m_1, \ldots, m_{3n}$. Let T be the metric in $\mathfrak{M}$ defined by the kinetic energy part of H. Then (as is shown by straightforward computation) the canonical mapping of $\mathcal{L}^2(E^{3n}, \alpha)$ (where α is Lebesgue measure) on $\mathcal{H}_{\mathfrak{M}}$ takes the formal dynamical operator associated with H in Sec. 2-4 into the operator $-\hbar\nabla^T + Q_V/\hbar$, and the operator defining the position observables into the Q_{q_i}. In other words the prescription for "quantizing" a classical system given in Sec. 2-4 may be reformulated to read as follows: Let $\mathfrak{M}$ be the configuration space manifold, let T be the kinetic energy metric, and let V be the potential energy function. Then (a) the Hilbert space of the corresponding quantum system is $\mathcal{H}_{\mathfrak{M}}$, (b) for each real-valued Borel function f on $\mathfrak{M}$ the corresponding quantum observable is that corresponding to the operator Q_f, and (c) the dynamical operator coincides with $-\hbar\nabla^T + Q_V/\hbar$ wherever this operator is defined.

This formulation is independent of any coordinate system and makes sense when $\mathfrak{M}$ is an arbitrary C_∞ manifold, i.e., for the dynamical systems considered in Sec. 1-3. We define the resulting quantum system to be the canonical quantization of the corresponding classical system. By arguments analogous to those used in the preceding section it can be shown to be the only quantization having certain simple desirable properties. Of course there remains the question of existence and uniqueness for self-adjoint extensions of the formal operator $-\hbar\nabla^T + Q_V/\hbar$. The remarks made in the preceding section on this subject still apply.

We remind the reader that in using non-rectangular coordinates one has an observable associated to a coordinate only when it is a *single*-valued *real*-valued function on the manifold of all configurations. Thus in using polar coordinates r, θ, ϕ for a single particle there will be no self-adjoint operator corresponding to θ or ϕ, as these are multiple-valued functions. On the other hand, there will be one corresponding to $\sin\theta$, $\cos\theta$, $\cos\phi$, $\sin\phi$, etc. Also there will be one corresponding to the *discontinuous* single-valued function obtained by insisting that θ be between 0 and 2π. In addition one can extend the notion of observable to include question-valued measures defined on manifolds other than the real line. When this is done, however, the new observables will not correspond to self-adjoint operators—indeed they need not correspond to operators of any kind. For example, one can associate with the angle θ a projection-valued measure defined on the circle $|z| = 1$ in the complex plane, and each state vector ϕ will define a probability distribution on $|z| = 1$. This probability distribution will be consistent with that defined in the usual way for the regular observables $\sin\theta$ and $\cos\theta$ by this state vector. Actually in this case one can associate a unitary operator with the observable. An extension of the spectral theorem sets up a one-to-one correspondence between projection-valued measures in the plane and so-called "normal" operators. The normal operators with spectrum on $|z| = 1$ are just the unitary ones.

A momentum observable in classical mechanics is a function g on $\mathfrak{M}_{V^*}$ such that $\widetilde{dg}$ is the infinitesimal generator of a one-parameter group on $\mathfrak{M}_{V^*}$ generated by a one-parameter group in $\mathfrak{M}$. This one-parameter group in $\mathfrak{M}$ generates a one-parameter unitary group in $\mathfrak{H}_{\mathfrak{M}}$, and $\hbar/i$ times its infinitesimal generator is by definition the operator associated with the corresponding momentum observable in quantum mechanics. This definition is a natural one and is further justified by the fact that equality between velocity and momentum observables in classical mechanics is preserved in quantum mechanics. The velocity observable associated with the position observable defined by Q_f is of course the observable associated with the operator $i[HQ_f - Q_fH]$.

Note that position and momentum observables are associated with rigorously defined self-adjoint operators in $\mathfrak{H}_{\mathfrak{M}}$. A dynamical operator H maps the Q_f for suitably differentiable f linearly onto a certain subset of the momentum observables and the kinetic part of H is determined (at least in a formal way) by this map. It would be interesting to have the formal part of these considerations worked up into a rigorous theory.

The intrinsic Hilbert space $\mathfrak{H}_{\mathfrak{M}}$ can be used for a convenient direct comparison of classical and quantum mechanics. We recall that whereas quantum mechanics is concerned with one-parameter groups of automorphisms of $\mathcal{H}_{\mathfrak{M}}$, classical mechanics is concerned with one-parameter groups of automorphisms of another object intrinsically defined on $\mathfrak{M}$—the cotangent bundle $\mathfrak{M}_{V^*}$. Now let $\phi = \alpha, e^{iS}$ be any element of $\mathcal{H}_{\mathfrak{M}}$ such that S is differentiable. S is arbitrary up to multiples of 2π but dS is uniquely defined by ϕ and is a one-to-one map of $\mathfrak{M}$ into $\mathfrak{M}_{V^*}$. Let $\beta_\phi(E) = \alpha(dS^{-1}(E))$ for all Borel sets E in $\mathfrak{M}_{V^*}$. Then β_ϕ is a finite measure in $\mathfrak{M}_{V^*}$ concentrated in $dS(\mathfrak{M})$. Moreover, β_ϕ uniquely determines ϕ. Thus $\phi \to \beta_\phi$ is a one-to-one map of a dense subspace of $\mathcal{H}_{\mathfrak{M}}$ into the set of all finite measures on $\mathfrak{M}_{V^*}$. Under the dynamical group in $\mathcal{H}_{\mathfrak{M}}$ the measures β_ϕ move in $\mathfrak{M}_{V^*}$. When β_ϕ is highly concentrated the movement in $\mathfrak{M}_{V^*}$ of the point which it approximates will be along a trajectory of the corresponding classical system. It would be interesting to have a rigorous proof of a precisely formulated theorem along these lines.

2-7 Linear Systems and the Quantization of the Electromagnetic Field

In Sec. 1-4 we showed how to rephrase classical mechanics in the linear case so that it applied to systems with an infinite number of degrees of freedom. In this section we shall show that the procedure for passing from a classical to a quantum system can be rephrased in the linear case so that it too makes sense for systems with an infinite number of degrees of freedom. We shall then show how the quantized electromagnetic field provides a model for radiation theory which synthesizes its wave and particle aspects.

We shall need to make extensive use (not only here but in the next chapter) of "tensor products" of Hilbert spaces and begin by outlining the pertinent facts about these products. Let $\mathcal{H}_1$ and $\mathcal{H}_2$ be Hilbert spaces. By an anti-bilinear function on $\mathcal{H}_1 \times \mathcal{H}_2$ we shall mean a complex-valued function f such that $f(a_1\phi_1 + a_2\phi_2, \theta) = \bar{a}_1 f(\phi_1, \theta) + \bar{a}_2 f(\phi_2, \theta)$ and $f(\phi, a_1\theta_1 + a_2\theta_2) = \bar{a} f(\phi, \theta_1) + \bar{a}_2 f(\phi, \theta_2)$, where a_1 and a_2 are complex numbers and $\phi, \phi_1, \phi_2, \theta, \theta_1, \theta_2$ are members of $\mathcal{H}_1$ and $\mathcal{H}_2$. If ϕ is in $\mathcal{H}_1$ and ψ is in $\mathcal{H}_2$ then $\phi \times \psi$ the *tensor product* of ϕ and ψ is defined to be the anti-bilinear function which takes θ_1, θ_2 into $(\phi \cdot \theta_1)(\psi \cdot \theta_2)$. Let $(\mathcal{H}_1 \otimes \mathcal{H}_2)'$ denote the set of all finite linear combinations of anti-bilinear functions of the form $\phi \times \psi$. It is an easy theorem whose proof we leave to the reader that there is one and only one way of introducing a positive definite inner product into $(\mathcal{H}_1 \otimes \mathcal{H}_2)'$ so that $(\phi_1 \times \psi_1) \cdot (\phi_2 \times \psi_2) = (\phi_1 \cdot \phi_2)(\psi_1, \psi_2)$. We call $(\mathcal{H}_1 \otimes \mathcal{H}_2)'$ equipped with this inner product the pretensor product of $\mathcal{H}_1$ and $\mathcal{H}_2$. Its completion is a Hilbert space which we call the *tensor product* of $\mathcal{H}_1$ and $\mathcal{H}_2$ and denote by $\mathcal{H}_1 \otimes \mathcal{H}_2$. It is not hard to show that norm convergence of a sequence of members of $(\mathcal{H}_1 \otimes \mathcal{H}_2)'$

implies pointwise convergence as functions on $\mathcal{H}_1 \times \mathcal{H}_2$ so that every member of $\mathcal{H}_1 \otimes \mathcal{H}_2$ may be identified with a certain anti-bilinear function on $\mathcal{H}_1 \times \mathcal{H}_2$. Let $\phi_1, \phi_2, \ldots$ be an orthonormal basis for $\mathcal{H}_1$ and let $\psi_1, \psi_2, \ldots$ be an orthonormal basis for $\mathcal{H}_2$. Then, as is easy to see, the $\phi_i \times \psi_j$ form an orthonormal basis for $\mathcal{H}_1 \otimes \mathcal{H}_2$. More generally, let $\mathcal{H}_1 = \mathcal{L}^2(\mathfrak{M}_1, \alpha_1)$, $\mathcal{H}_2 = \mathcal{L}^2(\mathfrak{M}_2, \alpha_2)$, $\mathcal{H}_3 = \mathcal{L}^2(\mathfrak{M}_1 \times \mathfrak{M}_2, \alpha_1 \times \alpha_2)$. If f is in $\mathcal{L}^2(\mathfrak{M}_1, \alpha_1)$ and g is in $\mathcal{L}^2(\mathfrak{M}_2, \alpha_2)$ then fg is in $\mathcal{L}^2(\mathfrak{M}_1 \times \mathfrak{M}_2, \alpha_1 \times \alpha_2)$. It is routine to show that there is a unitary map of $\mathcal{H}_1 \otimes \mathcal{H}_2$ on $\mathcal{H}_3$ which takes each $f \times g$ into fg.

Consider the case in which $\mathcal{H}_1 = \mathcal{H}_2 = \mathcal{H}$. The closed subspace of $\mathcal{H} \otimes \mathcal{H}$ generated by the elements of the form $\phi \times \psi + \psi \times \phi$ is easily seen to be just the set of all members of $\mathcal{H} \otimes \mathcal{H}$ which as functions on $\mathcal{H} \times \mathcal{H}$ are invariant under a switch of the arguments. We call this closed subspace the "symmetric tensor product" of $\mathcal{H}$ with itself and denote it by $\mathcal{H} \circledS \mathcal{H}$. In a similar fashion we define $\mathcal{H} \textcircled{A} \mathcal{H}$—the anti-symmetric tensor product—as the subspace generated by elements of the form $\phi \times \psi - \psi \times \phi$. It is easy to see that $\mathcal{H} \circledS \mathcal{H}$ and $\mathcal{H} \textcircled{A} \mathcal{H}$ are orthogonal complements of one another. In fact the operator that interchanges arguments is both unitary and self-adjoint and hence has a pure point spectrum with 1 and -1 as its only eigenvalues. $\mathcal{H} \circledS \mathcal{H}$ and $\mathcal{H} \textcircled{A} \mathcal{H}$ are the 1 and -1 eigenspaces, respectively.

By an obvious generalization of the preceding, we define anti-multilinear functions on $\mathcal{H}_1 \times \mathcal{H}_2 \times \cdots \times \mathcal{H}_n$ and a tensor product $\mathcal{H}_1 \otimes \mathcal{H}_2 \times \cdots \otimes \mathcal{H}_n$ which is the closed linear span of the products $\phi_1 \times \phi_2 \times \cdots \times \phi_n$ and in which

$$(\phi_1 \times \phi_2 \times \cdots \times \phi_n) \cdot (\theta_1 \times \theta_2 \times \cdots \times \theta_n) = (\phi_1 \cdot \theta_1)(\phi_2 \cdot \theta_2) \ldots (\phi_n \cdot \theta_n)$$

As before, we have a natural unitary mapping of $\mathcal{L}^2(\mathfrak{M}_1 \times \mathfrak{M}_2 \times \cdots \times \mathfrak{M}_n, \alpha_1 \times \alpha_2 \times \cdots \times \alpha_n)$ onto $\mathcal{L}^2(\mathfrak{M}_1, \alpha_1) \otimes \mathcal{L}^2(\mathfrak{M}_2, \alpha_2) \otimes \cdots \otimes \mathcal{L}^2(\mathfrak{M}_n, \alpha_n)$. We note in particular that the Hilbert space of states for a system of n particles is the tensor product of the Hilbert spaces of states for the individual particles. Each permutation of the n variables in $\mathcal{H} \otimes \mathcal{H} \otimes \cdots \otimes \mathcal{H}$ (n factors) induces a unitary transformation of this space. We denote the closed subspace of all vectors in $\mathcal{H} \otimes \mathcal{H} \otimes \cdots \otimes \mathcal{H}$ that are invariant under *all* of these unitary transformations by $\mathcal{H} \circledS \mathcal{H} \circledS \cdots \circledS \mathcal{H}$ and call it the symmetric tensor product of the $\mathcal{H}$'s. The closed subspace of all vectors that are carried into their negatives by every transformation defined by an odd permutation we call the "anti-symmetric tensor product" of the $\mathcal{H}$'s and denote by $\mathcal{H} \textcircled{A} \mathcal{H} \textcircled{A} \cdots \textcircled{A} \mathcal{H}$. It is *not* true that $\mathcal{H} \otimes \mathcal{H} \otimes \cdots \otimes \mathcal{H}$ is the direct sum of $\mathcal{H} \circledS \cdots \circledS \mathcal{H}$ and $\mathcal{H} \textcircled{A} \cdots \textcircled{A} \mathcal{H}$.

Let $T_1, T_2, \ldots, T_n$ be bounded linear operators in $\mathcal{H}_1, \mathcal{H}_2, \ldots, \mathcal{H}_n$. Then there exists a unique bounded linear operator T such that $T(\phi_1 \times \phi_2 \times \cdots \times \phi_n) = T_1(\phi_1) \times T_2(\phi_2) \times \cdots \times T_n(\phi_n)$. We denote it by $T_1 \times T_2 \times \cdots \times T_n$ and call it the tensor product of the T_j. If the T_j

are all unitary then so is their tensor product and similarly for self-adjointness. For unbounded self-adjoint operators one has to examine questions of domain, but these cause no real difficulties and one defines $T_1 \times T_2 \times \cdots \times T_n$ as an unbounded self-adjoint operator.

Let $t \rightarrow U_t^j$, $j = 1, 2, \ldots, n$ be one-parameter unitary groups in the Hilbert spaces $\mathcal{H}_1, \mathcal{H}_2, \ldots, \mathcal{H}_n$. Then $t \rightarrow U_t^1 \times U_t^2 \times \cdots \times U_t^n$ is a one-parameter unitary group in $\mathcal{H}_1 \otimes \mathcal{H}_2 \otimes \cdots \otimes \mathcal{H}_n$ which we denote by $U^1 \otimes U^2 \otimes \cdots \otimes U^n$. Let K^j denote the infinitesimal generator of U^j. Routine calculation then shows that $(K^1 \times I \times \cdots \times I) + (I \times K^2 \times I \times \cdots \times I) + (I \times I \times K^3 \times I \times \cdots \times I) + \cdots + (I \times \cdots \times I \times K^n)$ is the infinitesimal generator of $U^1 \otimes U^2 \otimes \cdots \otimes U^n$. A classical system consisting of n non-interacting particles has its quantum dynamical operator of this form.

If $\mathcal{H}_1 = \mathcal{H}_2 = \cdots = \mathcal{H}_n = \mathcal{H}$ and $U^1 = U^2 = \cdots = U^n = V$ then $V \otimes V \otimes \cdots \otimes V$ carries each of the subspaces $\mathcal{H} \circledS \mathcal{H} \circledS \cdots \circledS \mathcal{H}$ and $\mathcal{H} \textcircled{A} \mathcal{H} \textcircled{A} \cdots \textcircled{A} \mathcal{H}$ into itself and defines one-parameter unitary groups there. We shall call these the symmetric and anti-symmetric n-th powers of V and denote them by $V \circledS V \circledS \cdots \circledS V$ and $V \textcircled{A} V \textcircled{A} \cdots \textcircled{A} V$, respectively.

Let $\mathcal{H}_1, \mathcal{H}_2, \ldots$ be a sequence of Hilbert spaces and let $\mathcal{H}$ be the set of all sequences $\phi_1, \phi_2, \ldots$ such that

$$\sum_{j=1}^{\infty} \| \phi_j \|^2 < \infty$$

Then $\mathcal{H}$ is itself a Hilbert space in an obvious and natural way. We call the Hilbert space the direct sum of the $\mathcal{H}_i$ and denote it by $\mathcal{H}_1 \oplus \mathcal{H}_2 \oplus \cdots$. If U^j is a one-parameter unitary group in $\mathcal{H}_j$ then we define a one-parameter unitary group $U^1 \oplus U^2 \oplus \cdots$ in $\mathcal{H}_1 \oplus \mathcal{H}_2 \oplus \cdots$ as follows: $(U^1 \oplus U^2 \oplus \cdots)_t (\phi_1, \phi_2, \ldots) = U_t^1(\phi_1), U_t^2(\phi_2), \cdots$. We call $U^1 \oplus U^2 \oplus \cdots$ the direct sum of the U^j.

Now let us return to quantum mechanics. One of the main results of Sec. 1-4 was that for a *linear* classical system the phase space could be given the structure of a finite dimensional Hilbert space $\mathcal{H}_0$ in such a way that the dynamical group is a one-parameter unitary group V. Now when we quantize this linear system according to the principles enunciated in Secs. 2-4 and 2-5 we get an infinite dimensional Hilbert space $\mathcal{H}$ and a one-parameter unitary group U in $\mathcal{H}$. It is natural to ask whether there is any simple relationship between U and V, and we shall see that there is. Specifically, it can be shown that there is a unitary map of $\mathcal{H}$ on $C \oplus \mathcal{H}_0 \oplus \mathcal{H}_0 \circledS \mathcal{H}_0 \oplus \mathcal{H}_0 \circledS \mathcal{H}_0 \circledS \mathcal{H}_0 \oplus \cdots$ such that U goes into $I \oplus V \oplus V \circledS V \oplus \cdots$. Here C is the one-dimensional Hilbert space of all complex numbers and I is the identity one-parameter group of unitary operators in C.

Let $\mathcal{H}^r = \mathcal{H}_0 \oplus \mathcal{H}_1 \oplus \mathcal{H}_2 \oplus \cdots$ be any direct sum of finite or infinitely dimensional Hilbert spaces $\mathcal{H}_j$. If $X_0, X_1, X_2, \ldots$ is any sequence of bounded linear operators, where X_j has domain $\mathcal{H}_j$ and range in $\mathcal{H}_{j+1}$, then we define an operator $X = (X_0, X_1, X_2, \ldots)^r$ by setting

$$X(\phi_0, \phi_1, \phi_2, \ldots) = X_0^*(\phi_1), X_0(\phi_0) + X_1^*(\phi_2), X_1(\phi_1) + X_2^*(\phi_3), \ldots$$

whenever both sequences belong to $\mathcal{H}$. It is not hard to verify that X is self-adjoint. Now consider the special case in which $\mathcal{H}_0 = C$ and $\mathcal{H}_n = \mathcal{H}_1 \circledS \mathcal{H}_1 \circledS \cdots \circledS \mathcal{H}_1$ (n factors). For each ϕ in $\mathcal{H}_1$ let $X_n^\phi(\theta) = \sqrt{(n+1)/2}\ P_{n+1}(\phi \times \theta)$, where P_{n+1} denotes the projection of $\mathcal{H}_1 \otimes \mathcal{H}_1 \otimes \cdots \otimes \mathcal{H}_1$ $(n+1$ factors) onto its closed subspace $\mathcal{H}_1 \circledS \mathcal{H}_1 \circledS \cdots \circledS \mathcal{H}_1$ $(n+1$ factors), and let $X_\phi = (X_0^\phi, X_1^\phi, X_2^\phi, \ldots)^r$. A routine calculation shows that $X_{\phi_1} X_{\phi_2} - X_{\phi_2} X_{\phi_1}$ is equal to $[(\phi_2 \cdot \phi_1) - (\phi_1 \cdot \phi_2)]/2$ times the identity wherever it is defined and it is almost obvious that for every unitary operator W in $\mathcal{H}_1$ we have

$$X_{W(\phi)} = e^W X_\phi e^{-W}$$

where e^W is the unitary operator taking each $\mathcal{H}_n$ onto itself and in $\mathcal{H}_n$ is the restriction to $\mathcal{H} \circledS \cdots \circledS \mathcal{H}$ of $\phi_1 \cdots \phi_n \to W(\phi_1) \cdots W(\phi_n)$. In particular if $\mathcal{H}_1$ is the phase space of a classical, finite dimensional linear system made into a Hilbert space and $t \to V_t$ is the dynamical group then

$$X_{V_t(\phi)} = e^{V_t} X_\phi\ e^{-V_t}$$

Let $\mathcal{H}_1 = \mathfrak{M} \oplus \mathfrak{M}^*$, where the vector space $\mathfrak{M}$ is the configuration space of the classical system, and let T denote the linear mapping of $\mathfrak{M}$ on $\mathfrak{M}^*$ defined by the kinetic energy. Let B denote the operator introduced on p. 32 so that $\theta \to \frac{1}{2} T(B(\theta), (B(\theta))$ denotes the potential energy function on $\mathfrak{M}$. For each $\phi_1 = \theta_1, \ell_1$ in $\mathfrak{M} \dotplus \mathfrak{M}^*$, let $f_{\phi_1}(\theta, \ell) = \hbar(\ell_1(\theta) - \ell(B(\theta_1)))$. Then f_{ϕ_1} is linear on $\mathfrak{M} \dotplus \mathfrak{M}^*$ regarded as a real vector space, and every linear f on $\mathfrak{M} \dotplus \mathfrak{M}^*$ is uniquely an f_ϕ. Moreover the Poisson bracket $[f_{\phi_1}, f_{\phi_2}]$ is readily computed to be $\hbar[\ell_1(B(\theta_2) - \ell_2(B(\theta_1)]$ which is just $\hbar$ times the imaginary part of the complex inner product of ϕ_1 and ϕ_2 in $\mathcal{H}_1$. Hence $(X_{\phi_1} X_{\phi_2} - X_{\phi_2} X_{\phi_1}) = (\hbar/i)[f_{\phi_1}, f_{\phi_2}]$ and if we assign to each observable f_ϕ the self-adjoint operator X_ϕ we see that the Heisenberg commutation relations are satisfied. From $X_{V_t(\phi)} = e^{V_t} X_\phi e^{-V_t}$ we deduce that $(1/i)(HX_\phi - X_\phi H) = X_{A(\phi)}$, where $-iH$ is the skew-adjoint infinitesimal generator of $t \to e^{V_t}$. But a simple computation shows that $f_{A(\phi)} = A^*(f_\phi)$, which is the Poisson bracket of the classical energy with f_ϕ. Thus $\hbar H$ and the X_ϕ satisfy just the commutation relations satisfied by the operators in the canonical quantization of our classical linear system corresponding to the energy and the observables f_ϕ. Since the X_ϕ for the f_ϕ which vanish on $\mathfrak{M}$ are easily seen to be a complete commuting family we may appeal to the uniqueness theorem quoted in Sec. 2-4 and conclude that we may map $\mathcal{H}$, the Hilbert space of the classical quantization, onto $\mathcal{H}^r = C \oplus \mathcal{H}_1 \oplus \mathcal{H}_1 \circledS \mathcal{H}_1 \oplus \cdots$ in such a manner that the operator corresponding to the observable f_ϕ goes over into X_ϕ and the dynamical group into $e^V = I \oplus V \oplus V \circledS V \oplus V \circledS V \circledS V \oplus \cdots$. Otherwise formulated we have the following equivalent prescription for setting up the canonical quantization of a classical system when the system is linear. Convert the phase space into a finite dimensional Hilbert space $\mathcal{H}_1$ so that the dynamical group becomes a one-parameter unitary group $t \to V_t$ as described in Sec. 1-4. Form the one-parameter group $e^V = I \oplus V \oplus V \circledS V \oplus V \circledS V \circledS V \oplus \cdots$. This is the dynamical group. For each real linear functional f on $\mathcal{H}_1$ choose ϕ so that $f = f_\phi$. X_ϕ is the self-adjoint operator corresponding to the observable f.

Let $\nu_1, \nu_2, \ldots, \nu_k$ be the fundamental frequencies of the classical system. Then the eigenvalues of the dynamical operator which correspond to eigenfunctions in the subspace $\mathcal{H}(V)$ are just $2\pi\nu_1, 2\pi\nu_2, \ldots, 2\pi\nu_k$. Hence the corresponding eigenvalues of the energy operator are $2\pi\hbar\nu_1, 2\pi\hbar\nu_2, \ldots, 2\pi\hbar\nu_k$ or $h\nu_1, h\nu_2, \ldots, h\nu_k$ if we take $\hbar = h/2\pi$. Those lying in $\mathcal{H}(V) \circledS \mathcal{H}(V)$ are all possible sums $h(\nu_i + \nu_j)$, those in $\mathcal{H}(V) \circledS \mathcal{H}(V) \circledS \mathcal{H}(V)$ are all sums $h(\nu_i + \nu_j + \nu_k)$, etc., where there are multiplicities that the reader can compute without difficulty. It follows that the most general energy eigenvalue is of the form $h(n_1\nu_1 + n_2\nu_2 + \cdots + n_k\nu_k)$. The reader will find that these values differ by a constant from those that would be obtained by applying the canonical quantization directly. This is explained by the fact that the commutation relations only determine H up to an additive constant. We remind the reader that in general energy is determined only up to an additive constant in both classical and quantum mechanics. Our new form of the canonical quantization simply chooses this constant in a different way—a better way from the point of view of infinite dimensional generalization.

The importance of our new way of "quantizing" linear systems lies in the fact that all constructions apply equally well whether $\mathcal{H}_1$ is finite or infinite dimensional—with one exception: in the infinite dimensional case X_ϕ will be defined for only a dense set of f_ϕ's. Thus we have a natural way of quantizing an infinite dimensional, linear system. Given any such system we realize its dynamical group as a one-parameter unitary group in phase space made into a Hilbert space and proceed as above. This construction applies in particular when the classical system is the electromagnetic field. In this case—and in many similar cases—one introduces the observable associated with the unique self-adjoint operator N which is n times the identity in the subspace $\mathcal{H}(V) \circledS \mathcal{H}(V) \circledS \mathcal{H}(V) \circledS \cdots \circledS \mathcal{H}(V)$ (n factors) and calls it the "number of particles" observable. Furthermore one speaks of a state defined by a vector in $\mathcal{H}(V) \circledS \cdots \circledS \mathcal{H}(V)$ as one in which "are n particles." Why is this? In what sense does the system in such a state behave as though it were made up of n particles?

We saw on pages 82 and 93 that the operators corresponding to the energy and momentum observables of a quantum system are obtainable as the infinitesimal generators of certain one-parameter groups of unitary operators. When the system is that of a free particle (page 96) much more is true. Let $\mathcal{E}$ denote the group of all rigid motions in space, let $\mathcal{T}$ denote the group of all translations in time, and let $\tilde{\mathcal{E}}$ be the group of all transformations of space-time generated by $\mathcal{E}$, $\mathcal{T}$, and the transformations $x,y,z,t \to x - ut, y - vt, z - wt, t$. There is a natural mapping $\alpha \to U_\alpha$ of $\tilde{\mathcal{E}}$ into the unitary operators in the Hilbert space of our system such that for all α and β in $\mathcal{E}$, $U_\alpha U_\beta U_{\alpha\beta}^{-1}$ is a multiple of the identity and such that the position observables as well as the energy and momentum observables are obtainable as the infinitesimal generators of certain one-parameter subgroups of $\tilde{\mathcal{E}}$. In this sense the quantum mechanics of a free particle is completely described by the mapping $\alpha \to U_\alpha$, which (see page 127) is an example of what is called a projective unitary representation of $\tilde{\mathcal{E}}$.

Now, according to the special theory of relativity, the group of symmetries of space-time is not $\tilde{\mathcal{E}}$ but another group $\mathcal{L}$ of the same dimension (containing $\mathcal{E}$ and $\mathcal{T}$, and called the Poincaré group or the inhomogeneous Lorentz group. One can modify the quantum mechanics of a free particle so as to make it consistent with the demands of the special theory of relativity, and when this is done one finds that the resulting quantum system is completely described by a projective representation of $\mathcal{L}$. The projective representations of $\mathcal{L}$ which arise in this way turn out to be part of a larger system of representations with very similar properties.

These representations are found to be associated with the observables of quantum systems and, because of the similarity to systems describing relativized free particles, it is natural and fruitful to think of these systems as describing generalized particles. These generalized particles differ from those already described chiefly in having "internal degrees of freedom"; there may be observables which commute with all the linear momenta and yet are not functions of these (see the discussion of "spin" in Sec. 3-4). However a few of them differ somewhat more from classical particles in that they have zero (rest) mass and no position observables. In any event Maxwell's equations are Lorentz-invariant, so there is a natural "action" of $\mathcal{L}$ on the Hilbert space $C \oplus \mathcal{H}(V) \oplus \cdots$ which gives us a representation $\alpha \to U_\alpha$ of $\mathcal{S}$ by unitary operators in this Hilbert space. $\mathcal{H}(V)$ is invariant under the U_α, and the "subrepresentation" of $\mathcal{L}$ so-defined turns out to be one of those describing a generalized free particle. Moreover each summand $\mathcal{H}(V) \circledS \mathcal{H}(V) \circledS \cdots \circledS \mathcal{H}(V)$ is invariant under U and the action is that induced by the action in $\mathcal{H}(V)$. Now observe that when $\mathcal{H}(V)$ is the Hilbert space of the quantum system of a particle then $\mathcal{H}(V) \otimes \mathcal{H}(V)$ is the Hilbert space of the quantum system of two independent such particles and the dynamical group is $V \otimes V$. Similarly $\mathcal{H}(V) \otimes \mathcal{H}(V) \otimes \cdots \mathcal{H}(V)$ (n factors) is the Hilbert space of the quantum system of n independent such particles. Hence if it were not for the symmetric reduction—if we had $\mathcal{H}(V) \otimes \mathcal{H}(V) \cdots \otimes \mathcal{H}(V)$ instead of $\mathcal{H}(V) \circledS \mathcal{H}(V) \cdots \circledS \mathcal{H}(V)$—there would be a clear and definite sense in which each state defined by a vector in the n eigenspace of N is a state in which there are n independent particles each moving according to the quantum laws implied by the action of the U_α in $\mathcal{H}(V)$. What is the significance of the symmetric reduction? Quite apart from the quantization of the electromagnetic or other field, what does it mean when in setting up the quantum mechanics of n independent identical particles we find that the only states occurring are those in the symmetric subspace of the tensor product to which we are led by the canonical quantization? Since this subspace is invariant under the dynamical group we still have a well-defined quantum system. However there are many missing observables. Only those self-adjoint operators that carry the symmetric subspace into itself can correspond to observables. This means in particular that there is no observable for the x coordinate of any particular particle but only for symmetric functions of the x coordinates of all of the particles. Now, of course, knowing $x_1 + x_2 + \cdots + x_n$, $x_1^2 + x_2^2 + \cdots + x_n^2$, $x_1^3 + x_2^3 + \cdots + x_n^3, \cdots, x_1^n + x_2^n + \cdots + x_n^n$ we can use algebra to find $x_1, x_2, \ldots, x_n$ *but not* which is which. For example, suppose $n = 2$. If we know that $x_1 + x_2 = a$ and $x_1^2 + x_2^2 = b$, then $x_1 x_2 = (a^2 - b)/2$ and so we know that x_1 and x_2 are the two roots of the equation $x^2 - ax + (a^2 - b)/2 = 0$. However there is no way of telling which root is x_1 and which is x_2. The accepted physical interpretation of such a state of affairs is that the particles are rather far from being like billiard balls in that they have no individual identities—interchanging the places of two of them makes no physical difference whatever and it makes no sense to talk about the x coordinate of a particular one of them. An analogy may help to make the situation clear. Imagine a vibrating string stretched between two supports and in the state of motion indicated in Fig. 1, where the arrows point in the direction of motion of the disturbances: After a time the two disturbances will pass one another and the string will be as indicated in

Fig. 1

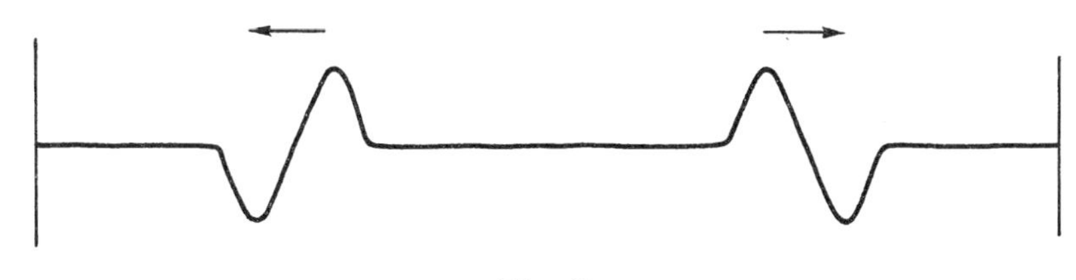

Fig. 2

Fig. 2. Still later they will be reflected at the supports, bounce back, and the string will be as in Fig. 3. If we have kept our eyes on the two disturbances, regarded as individual entities they seem to have

Fig. 3

changed places between Fig. 1 and Fig. 3. On the other hand, the string itself is in exactly the same state in Fig. 1 as it is in Fig. 3.

It turns out that all of the "elementary particles" of physics have this character of indistinguishability and are to be thought of as more like the disturbances in a string than like individual classical particles. In many cases, however, the identity manifests itself through the appearance of the anti-symmetric subspace of the n-fold tensor product instead of the symmetric one. In the symmetric case one says that the particles obey Bose-Einstein statistics and calls them bosons. In the anti-symmetric case, one speaks of Fermi-Dirac statistics and of fermions.

To return to the quantized electromagnetic field we see that we can look upon it equivalently as a collection of an indefinite number of independent indistinguishable particles that obey Bose-Einstein statistics. These particles are called *photons*. We emphasize that the quantized system is at the *same time* a quantized vibrating continuum and a quantized collection of particles. It all depends upon which observables are looked at. The mathematical system comprises both the wave and particle aspects of electromagnetic radiation, and there is a paradox only when one tries to frame too naïve a physical picture.

We conclude by remarking that photons differ from classical particles in two other respects. In the classical limit they do not obey

Newton's laws but travel always with the velocity of light. Also there are non-classical observables—the spin angular momenta. We shall discuss these in more detail in the next chapter—but when the particles are electrons.

2-8 Quantum-Statistical Mechanics

The statistical theory of heat and temperature discussed in Sec. 1-5 extends rather completely to quantum mechanics. To find the quantum analog of the Gibbs canonical state we seek a mixed state defined by a non-negative self-adjoint trace operator Γ which represents minimal knowledge or maximum "randomness" subject to the condition that the energy observable H_0 have a given expected value E. As we have seen, the expected value in question is $\text{Trace}(\Gamma H_0)$. To measure the "amount of randomness" in Γ we again appeal to information theory. If $\phi_1, \phi_2, \ldots$ is a complete orthonormal set of eigenvectors for Γ, where $\Gamma(\phi_j) = \gamma_j \phi_j$, then $\gamma_1 + \gamma_2 + \cdots = 1$ and the state defined by Γ is $\gamma_1 \alpha_{\phi_1} + \gamma_2 \alpha_{\phi_2} + \cdots$ where α_{ϕ_j} is the pure state defined by ϕ_j. Taking pure states as representing zero randomness, information theory suggests that $\Sigma \gamma_j \log 1/\gamma_j$ be taken as the measure of randomness of Γ. Now by the operational calculus we may form the operator $\Gamma \log \Gamma$ and we see that $\Gamma \log \Gamma(\phi_j) = (\gamma_j \log \gamma_j)\phi_j$. Hence $\Sigma \gamma_j \log 1/\gamma_j = -\text{Trace}(\Gamma \log \Gamma)$. In other words, we seek a positive trace operator Γ such that $-\text{Trace}(\Gamma \log \Gamma)$ is as large as possible subject to the side condition, $\text{Trace}(\Gamma H_0) = E$. Using techniques from the calculus of variations, adapted to operators (cf. von Neumann's book), it can be shown that Γ must be of the form $Ae^{-H_0/B}$ where A and B are constants. Since $\text{Trace}\,\Gamma = 1$ we must have $A = [\text{Trace}(e^{-H_0/B})]^{-1}$ so that Γ must be of the form $e^{-H_0/B}/\text{Trace}(e^{-H_0/B})$. Of course for many choices of H_0 these conditions will be impossible of fulfillment. However in the systems of interest in this connection H_0 has the following properties:

(a) H_0 has a pure point spectrum and all eigenvalues are non-negative.

(b) $P(B) = \text{Trace}(e^{-H_0/B})$ exists for all B greater than zero.

(c) $\text{Trace}(H_0 e^{-H_0/B})$ exists for all B greater than zero.

(d) $\text{Trace}(H_0 e^{-H_0/B})/P(B)$ is an unbounded monotone function of B.

For such an H_0 if we set $E(B) = \text{Trace}(H_0 e^{-H_0/B})/P(B)$ we see that for each E there is just one B such that the operator $e^{-H_0/B}/P(B)$ satisfies the required condition. The (mixed) state defined by this operator is the quantum analog of the Gibbs canonical state for the given value of E.

We can make our formulas more analogous to the classical ones by writing our traces in terms of the eigenvalues $\lambda_1, \lambda_2, \ldots$ of H_0. Let us list these so that multiple eigenvalues are repeated and $\lambda_1 \leq \lambda_2 \leq \cdots$. We can do this because

$$\sum_{j=1}^{\infty} e^{-\lambda_j/B}$$

converges. Then

$$P(B) = \sum_{j=1}^{\infty} e^{-\lambda_j/B} = \int_0^{\infty} e^{-x/B} \, d\beta_q(x)$$

where β_q is the measure which assigns to each Borel set the number of eigenvalues (counting multiplicities) that it contains. In particular $\beta_q([0, x])$ is the number of eigenvalues of H_0 that are less than or equal to x. Similarly, $E(B) = \int_0^{\infty} xe^{-x/B} \, d\beta_q(x)/P(B)$ and just as in the classical case $E(B) = B^2 P'(B)/P(B)$. We have a theory just like that in the classical case with one important exception. The continuous measure β [where $\beta([0, x])$ is the measure in phase space of the set where the classical Hamiltonian is less than or equal to x] is replaced by the discrete measure β_q, where $\beta_q([0, x])$ is the number of eigenvalues less than or equal to x. This reminds us of the arbitrary substitution of a continuous measure by a discrete one which we made in order to get Planck's radiation law. We now have a theoretical justification for this substitution. Indeed, as we saw in Sec. 2-7, the eigenvalues of H_0 for the case of the harmonic oscillator are just the values $(2k+1)\pi\hbar\nu$, where ν is the frequency of the oscillator and $k = 1, 2, \ldots$. Normalizing by subtracting the constant $\pi\hbar\nu$ from H_0 we get $2\pi\hbar\nu$, $4\pi\hbar\nu, \ldots$. Thus β_q assigns the measure 1 to each point of the form $2k\pi\hbar\nu$ and 0 to any set containing none of these points. Hence if we suppose once again that $\hbar = h/2\pi$ we find that β_q is $1/h$ times the measure β_h introduced on p. 58. Since multiplying the measure by a constant does not change $E(B)$ we are led by general quantum-mechanical principles to exactly the formula found on p. 58. Thus quantum mechanics solves the problem with which the old quantum theory began. We note that in quantum mechanics there is also no difficulty about the absence of a Liouville measure in the phase space of our infinite dimensional system. All systems, finite or infinite dimensional, have as phase space the lattice of closed subspaces of a separable, infinite dimensional Hilbert space, and the analog of the Liouville measure is the measure that assigns to each closed subspace its dimension.

It is interesting in the general case to compare the measure β_q provided by quantum mechanics with the measure β provided by classical mechanics. Since it is found that classical statistical mechanics agrees with experiment for high temperatures but not for low ones, it is to be expected that β and β_q will lead to the same specific heat formulas at high temperatures. From this and the theory of the Laplace transform $(\alpha \to \int_0^\infty e^{-xt}\, d\alpha(t))$ one is led to suppose that $\beta_q([0, x])$ is asymptotically a multiple of $\beta([0, x])$ for large x. That is, $\lim_{x \to \infty} \beta([0, x])/\beta_q([0, x])$ exists. This turns out to be the case. There are theorems in analysis which provide asymptotic formulas for the eigenvalue distribution of differential operators. When these are applied to the Schrödinger operator we find that

$$\lim_{x \to \infty} \frac{\beta([0, x])}{\beta_q([0, x])} = (2\pi\hbar)^N$$

where N is the dimension of configuration space. Thus, on the one hand, we have a theoretical explanation of the adequacy of classical statistical mechanics at high temperatures and, on the other, a physical motivation for the theorems about asymptotic eigenvalue distribution.

Chapter 3

GROUP THEORY AND THE QUANTUM MECHANICS OF THE ATOM

3-1 Preliminaries

It is natural to seek theoretical explanations of the spectra and chemical properties of atoms by applying the canonical quantization described in Sec. 2-4 to the classical mechanical system provided by the Rutherford model. This turns out to be wrong for two different reasons. There are at the same time more and fewer stationary states than are expected on this basis. One of the chief aims of this chapter is to explain the nature and significance of the changes that have to be made. Moreover, whether one uses the canonical quantization or the corrected one the mathematical analysis of the resulting quantum system presents formidable difficulties—analogous in some respects to those presented by the n-body problem in classical mechanics. We shall begin the chapter by discussing two general methods for obtaining some insight into the energy eigenvalue structure of complicated quantum systems. In one of these, one exploits symmetries of the system via the theory of group representations. In the other, one regards the given system as a "small perturbation" of a simpler system whose eigenvalues are known and uses power-series expansions in the perturbation parameter. The two methods are often used in conjunction. In studying the atom, for example, one can approximate it by a system in which the electrons do not interact and move in a common spherically symmetric field. This system is replete with symmetries and can be analyzed very completely. The actual system can then be discussed in terms of the approximate one by the power-series expansions of perturbation theory.

3-2 Basic Notions in the Theory of Group Representations

Let G be a locally compact topological group, that is, a group together with a topology under which it is locally compact and the group operations are continuous. A *continuous unitary representation* L of

G is by definition an assignment $x \to L_x$ of a unitary operator in some fixed Hilbert space $\mathcal{H}(L)$ to each x in G in such a manner that $L_{xy} = L_x L_y$ for all x and y in G and $x \to L_x(\phi)$ is continuous from G to $\mathcal{H}(L)$ for each fixed ϕ in $\mathcal{H}(L)$. It turns out that this last condition is implied by the apparently much weaker one that $x \to (L_x(\phi) \cdot \psi)$ is a Borel function for each fixed ψ and ϕ in $\mathcal{H}(L)$. Since we shall talk about no other kind it will be convenient to abbreviate the term "continuous unitary representation" to "representation." If G is the group of all rigid motions in E^3, f is in $\mathcal{L}^2(E^3)$, and for each α in G, we set $(L_\alpha f)(x, y, z) = f(\alpha(x, y, z))$, then L_α is unitary and $\alpha \to L_\alpha$ is a representation of G. Of course the continuous one-parameter unitary groups we have been discussing are just representations of the additive group of the real line.

Again let G be arbitrary. If L and M are representations of G and there exists a unitary operator U mapping $\mathcal{H}(L)$ on $\mathcal{H}(M)$ in such a fashion that $UL_xU^{-1} = M_x$ for all x in G then L and M are said to be *equivalent*. For arbitrary L and M we can construct a new representation $L \oplus M$ called the *direct sum* of L and M as follows: $\mathcal{H}(L \oplus M) = \mathcal{H}(L) \oplus \mathcal{H}(M)$ and $(L \oplus M)_x(\phi_1, \phi_2) = L_x(\phi_1), M_x(\phi_2)$. The set of all $\phi_1, 0$ is a closed subspace of $\mathcal{H}(L \oplus M)$ and this subspace is clearly invariant under all the operators $(L \oplus M)_x$. More generally, if H_1 is a closed subspace of $\mathcal{H}(L)$ which is such that $L_x(H_1)$ is contained in H_1 for all x then the L_x restricted to H_1 define a new representation L^{H_1} of G whose Hilbert space is H_1. We call L^{H_1} the *subrepresentation* of L defined by the closed invariant subspace H_1. It is almost trivial to show that if H_1 is closed and invariant then so is $H_1^\perp$, the orthogonal complement of H_1 in $\mathcal{H}(L)$. Moreover it is essentially obvious that L is equivalent to $L^{H_1} \oplus L^{H_1^\perp}$. Thus one can attempt to analyze a given representation L by writing it as a direct sum of certain subrepresentations. The notion of direct sum has an obvious generalization in which one allows several and even infinitely many direct summands. When L has no proper closed invariant subspaces or equivalently cannot be written as a direct sum of non-zero subrepresentations it is said to be *irreducible*. Ideally one would like to write every representation as a direct sum of (perhaps infinitely many) irreducible representations. This goal is attainable only in special cases. In general the situation is analogous to that which obtains when one has a self-adjoint operator without a pure point spectrum. Indeed a representation of the real line is a direct sum of irreducibles if and only if its self-adjoint infinitesimal generator has a pure point spectrum. On the other hand, by the celebrated Peter-Weyl theorem every representation of a *compact* G is a direct sum of *finite dimensional*, irreducible representations. We shall deal mainly with compact groups and make no attempt to deal with "continuous" decompositions. In general a representation which can be decomposed as a direct sum of irreducible ones will be called *discrete*. When one has a discrete representation it is natural to inquire into the extent to which the decomposition is unique. This question is easily answered by making use of a generalization of a fundamental lemma of Schur. If we rewrite the identity $UL_xU^{-1} = M_x$ defining the notion of equivalence in the form $UL_x = M_xU$ it makes sense for arbitrary bounded operators U from $\mathcal{H}(L)$ to $\mathcal{H}(M)$. We shall call a bounded operator from $\mathcal{H}(L)$ to $\mathcal{H}(M)$ that satisfies this identity an intertwining operator for L and M. The vector space of all intertwining operators for L and M will be denoted by $R(L, M)$. If $L = M$ then $R(L, M) = R(L, L)$. We write $R(L, L) = R(L)$ and call it the commutating algebra of L. Let L and M be arbitrary representations and let T be an element of $R(L, M)$. Then the null space N_T of T is trivially seen to be a closed invariant subspace of (L), and $\overline{R}_T$ the closure of the range of T is equally trivially seen to be a closed invariant subspace of $\mathcal{H}(M)$. Moreover T restricted to $N_T^{\perp}$ is a member T' of $R(L^{N_T^{\perp}}, M^{\overline{R}_T})$ which is one-to-one and has a dense range. It follows from a general theorem in operator theory that T' can be factored into the product of a self-adjoint operator in $R(L^{N_T^{\perp}})$ and a unitary operator from $N_T^{\perp}$ onto $\overline{R}_T$. This unitary operator sets up an equivalence between $L^{N_T^{\perp}}$ and $M^{\overline{R}_T}$. Thus we have:

Schur's lemma: If T is in $R(L, M)$ then L restricted to the orthogonal complement of the null space of T is equivalent to M restricted to the closure of the range of T.

Corollary 1: $R(L, M)$ reduces to zero if and only if no subrepresentation of L is equivalent to any subrepresentation of M. We then say that L and M are *disjoint*.

Corollary 2: If L and M are irreducible then $R(L, M) = 0$ if and only if L and M are not equivalent.

Suppose that $R(L)$ contains members other than constant multiples of the identity. Then since it contains T^* whenever it contains T it contains non-constant self-adjoint operators. Hence by the spectral theorem it contains non-trivial projections. But if $R(L)$ contains a projection the range of the projection is invariant under L. Hence we have:

Corollary 3: L is irreducible if and only if $R(L)$ consists only of scalar multiples of the identity.

Suppose that L and M are *primary* representations in the sense that they are direct sums of mutually equivalent irreducible representations: $L = L_1 \oplus L_2 \oplus \cdots$, $M = M_1 \oplus M_2 \oplus \cdots$. If L and M are equivalent or, more generally, are not disjoint, then there exists a non-zero member T of $R(L, M)$. T cannot reduce to zero on every $\mathcal{H}(L_j)$. Thus restricting to a suitable $\mathcal{H}(L_j)$ and projecting on a suitable $\mathcal{H}(M_k)$ we get a non-trivial member of $R(L_j, M_k)$. Thus L_j and M_k are equivalent. Thus in a primary representation the irreducible constituent is uniquely determined to within equivalence and two primary representations are disjoint if their irreducible constituents are not equivalent. It can also be shown—though we shall omit the proof—that two equivalent primary representations have the same number of irreducible constituents.

Now let L be any discrete representation. By grouping equivalent irreducible constituents together we may write it as $L_1 \oplus L_2 \oplus \cdots$ where the L_j are disjoint primary representations. Let $M = M_1 \oplus M_2 \oplus \cdots$ be any other direct sum of disjoint primary representations. Let U set up an equivalence between L and M. By an obvious adaptation of the above argument U must map $\mathcal{H}(L_j)$ *onto* some $\mathcal{H}(M_k)$ and set up an equivalence between L_j and M_k. Moreover if $L = M$ then every member of $R(L)$ must map each $\mathcal{H}(L_j)$ into itself. It follows at once that the $\mathcal{H}(L_j)$ are *uniquely determined subspaces* of $\mathcal{H}(L)$. Thus a discrete representation is uniquely a direct sum of discrete primary representations and every discrete primary representation is (non-uniquely) a direct sum of mutually equivalent irreducible representations. The irreducible summands *are unique* to within equivalence and the multiplicity with which they occur is uniquely determined. A fact which we wish to emphasize is that in the unique decomposition into primary subrepresentations each invariant subspace is carried into itself by every member of $R(L, L)$.

It follows from the above that one knows all discrete representations of a group to within equivalence when one knows all irreducible representations to within equivalence. In particular when the group is compact one knows all representations when one knows all irreducible ones (again to within equivalence of course). The problem of finding all irreducible representations of a given group is in general quite difficult. However there are many cases in which it has been solved. When the group is Abelian it follows at once from Corollary 3 above that all irreducible representations are one dimensional and have the form $x \to \chi(x)I$, where I is the identity and χ is a continuous homomorphism of the group into the group of complex numbers of modulus 1. If the group is the real line, the most general such homomorphism is $x \to e^{ixy}$, where y is a fixed real number.

An example, which will be of importance for us, in which all irreducible representations are known, is the compact group of all rotations about 0 in three-dimensional space E^3. For each $j = 0, 1, 2, \ldots$ let X_j denote the space of all complex-valued functions defined on the surface S^2 of the unit sphere in E^3 which are restrictions to S^2 of homogeneous polynomials of the j-th degree which satisfy Laplace's equation. It is easily seen that X_j is $2j + 1$ dimensional and is a closed subspace of the Hilbert space $\mathcal{L}^2(S^2)$. Moreover under $f(x, y, z) \to f(\alpha(x, y, z))$, where α is a rotation about 0, X_j is carried into itself and the transformation is unitary. Thus we define a $2j + 1$-dimensional unitary representation of the rotation group which we denote by D_j. It can be shown that each D_j is irreducible and that every irreducible unitary representation of the rotation group is equivalent to some D_j.

Now let K be a closed subgroup of our locally compact group G. If L is an irreducible representation of G we may restrict it to K and obtain a representation of K. This representation will not in general be irreducible and if discrete we may ask: Which irreducible representations of K will occur in its decomposition and with what multiplicities? Two cases of particular interest for quantum mechanics are the following: (a) G is the rotation group and K is the subgroup of rotations about a particular axis. (b) G is the direct product of the rotation group with itself and K is the "diagonal subgroup," the subgroup of all x, y with x = y. Consider case (a). Then K is isomorphic to the group of complex numbers of modulus 1 and the most general homomorphism χ of K is of the form $e^{i\theta} \to e^{im\theta}$, where m is an integer. Let us denote the representation $e^{i\theta} \to e^{im\theta}I$ by L_m. Then it is not difficult to show that D_j restricted to K is equivalent to the direct sum $L_{-j} \oplus L_{-j+1} \oplus \cdots L_{-1} \oplus L_0 \oplus L_1 \oplus \cdots \oplus L_j$.

Before considering case (b) we make some general remarks. If L and M are irreducible representations of G_1 and G_2, respectively, then $x, y \to L_x \times M_y$ is a representation of $G_1 \times G_2$ whose space is $\mathcal{H}(L) \otimes \mathcal{H}(M)$. We denote it by $L \times M$ and call it the outer Kronecker product of L and M. It is always irreducible and for a large class of groups, including the compact ones, it can be shown that every irreducible representation of $G_1 \times G_2$ is uniquely of the form $L \times M$, where L and M are irreducible representations of G_1 and G_2, respectively. If $G_1 = G_2 = G$ we may form the diagonal subgroup $\widetilde{G}$ of $G_1 \times G_2 = G \times G$ and consider the restriction of $L \times M$ to $\widetilde{G}$. Via the natural isomorphism of $\widetilde{G}$ with G we get a representation of G which in general is reducible. We denote it by $L \otimes M$ and call it the (inner) Kronecker product of L and M. Once we have determined the irreducible representations of G we may ask for the decomposition of $L \times M$ for each pair L and M and thus obtain a sort of "multiplication table" for the group. Case (b) of the problem mentioned above is clearly equivalent to determining the multiplication table for the rotation group. There is an easy answer which we state without proof:

$$D_j \otimes D_k = D_{|j-k|} \oplus D_{|j-k|+1} \oplus \cdots \oplus D_{j+k}$$

This formula is known as the Clebsch-Gordan series. Iterating it we can find the decomposition, or restriction to the diagonal, of an arbitrary irreducible representation of $G \times G \times \cdots \times G$ (n factors), where G is the rotation group.

3-3 Perturbations and the Group Theoretical Classification of Eigenvalues

Let U be a representation of the compact group G and let $L^1, L^2, L^3, \ldots$ be a set of inequivalent irreducible representations of G which has a member equivalent to every irreducible representation of G. Let H be a self-adjoint operator in $\mathcal{H}(U)$ which commutes with all U_x and has pure point spectrum. Let $\mathcal{H} = \mathcal{H}_1 \oplus \mathcal{H}_2 \oplus \mathcal{H}_3 \oplus \cdots$ be the decomposition of $\mathcal{H}$ induced by the unique primary decomposition of U, $\mathcal{H}_j$ being the space of a direct sum of replicas of $\mathcal{H}_j$. Let $\lambda_1, \lambda_2, \ldots$ be the eigenvalues of H and let H_{λ_j} denote the subspace of all eigenelements with eigenvalue λ_j. Since each U_x commutes with H it commutes with all P_{λ_j}, where P_{λ_j} is the projection on $\mathcal{H}_{\lambda_j}$ and hence carries each $\mathcal{H}_{\lambda_j}$ into itself. Hence each $\mathcal{H}_{\lambda_j}$ defines a subrepresentation of U and the decomposition of this into primary representations is implemented by the decomposition,

$$\mathcal{H}_{\lambda_j} = \mathcal{H}_{\lambda_j} \cap \mathcal{H}_1 \oplus \mathcal{H}_{\lambda_j} \cap \mathcal{H}_2 \oplus \cdots$$

Thus the $\mathcal{H}_{\lambda_j} \cap \mathcal{H}_k$ are mutually orthogonal subspaces that add up to $\mathcal{H}$. Moreover each $\mathcal{H}_{\lambda_j} \cap \mathcal{H}_k$ is both an eigenspace of H and the space of a primary subrepresentation of U. It might be noted that $x, t \to e^{iHt} U_x$ is a representation of $G \times R$, where R is the additive group of the line, and that $\mathcal{H}_{\lambda_j} \cap \mathcal{H}_k$ is the subspace of the primary subrepresentation defined by the irreducible representation $x, t \to e^{i\lambda_j t} L_x^k$. It follows that each $\mathcal{H}_{\lambda_j} \cap \mathcal{H}_k$ has a dimension which if not 0 or ∞ is a positive integral multiple of the dimension d_k of $\mathcal{H}(L^k)$ Thus whenever U contains irreducible components that are not one dimensional some eigenspaces of H are forced to be greater than one dimensional. The occurrence of multiple eigenvalues is referred to in quantum mechanics as degeneracy. We see then that symmetry of H, in the sense of being left invariant by a compact unitary group, tends to cause degeneracy. When each $\mathcal{H}_{\lambda_j}$ intersects just one $\mathcal{H}_k$ and when this intersection has the dimension of $\mathcal{H}(L^k)$ so that H has no more multiple eigenvalues than are forced upon it by its being in $R(L)$, one says that H has (relative to U) no *accidental* degeneracies. When H has no accidental degeneracies the eigenvalues are classified by the irreducible representations in the sense that each eigenspace is also the space of a unique L_j. In one important case which we shall encounter there are accidental degeneracies but only because there are several different $\mathcal{H}_k$'s which intersect a given $\mathcal{H}_{\lambda_j}$. When this is the case, i.e., when each $\mathcal{H}_{\lambda_j} \cap \mathcal{H}_k$ has dimension 0 or $\mathcal{H}(L^k)$ or, equivalently, when the primary components of $x, t \to e^{iHt} U_x$ are all irreducible we shall say that there are no *serious degeneracies*.

When H is the energy observable of a quantum-mechanical system one observes degeneracy by changing the system slightly so as to destroy the symmetry. Each λ_j then splits into as many distinct values as the dimension of $\mathcal{H}_{\lambda_j}$. One can also remove only part of the symmetry so that the altered H commutes with the U_x for x in some closed subgroup K of G. In this case λ_j splits into fewer parts than the dimension of $\mathcal{H}_{\lambda_j}$ and the nature of the splitting is closely related to the decomposition of the restriction of the associated L^k to K. We conclude the section with a detailed study of this relationship.

Let G, K, H, and U be as above and let H^ϵ be self-adjoint for $0 \leq \epsilon \leq 1$ and such that $H^0 = H$. We suppose that the dependence of H^ϵ on ϵ is sufficiently regular to justify the analytical steps to be taken below and in addition that $U_x H^\epsilon = H^\epsilon U_x$ for all x in K and all ϵ in $[0,1]$.

Let λ_j be an eigenvalue of H^0 and suppose that there are no serious degeneracies so that U restricted to $\mathcal{H}_{\lambda_j}$ is a direct sum of distinct irreducible representations. Let L^k be one of these and let $P_{j,k}$ denote the projection of $\mathcal{H}$ on $\mathcal{H}_{\lambda_j} \cap \mathcal{H}_k$. Let $H^\epsilon - H^0 = W^\epsilon$ and let $W^\epsilon_{j,k} = P_{j,k} W^\epsilon P_{j,k}$. [In this discussion we shall ignore the difficulties arising in forming sums and differences of unbounded operators.] Then $W^\epsilon_{j,k}$ is a self-adjoint operator in the finite dimensional Hilbert space $\mathcal{H}_{\lambda_j} \cap \mathcal{H}_k$. Its significance is that the eigenvalues of $d/d\epsilon\, [W^\epsilon_{j,k}]_{\epsilon=0}$ provide the coefficients of the first-order terms in the power series expansion of the eigenvalues of H^ϵ which for $\epsilon = 0$ reduce to eigenvalues associated with eigenvectors in $\mathcal{H}_{\lambda_j} \cap \mathcal{H}_k$. The fact that this is so is the fundamental assertion of first-order perturbation theory. Leaving rigor aside let us see how this comes about. Let ϕ^ϵ be a vector in $\mathcal{H}$ depending analytically upon ϵ so that we may write $\phi^\epsilon = \phi^0 + \epsilon\phi_1 + \epsilon^2\phi_2 + \cdots$, where ϕ^0 is in $\mathcal{H}_{\lambda_j} \cap \mathcal{H}_k$, and that ϕ^ϵ is an eigenvector of H^ϵ with eigenvalue $\lambda^\epsilon = \lambda_j + a_1\epsilon + a_2\epsilon^2 + \cdots$. Then

$$[H^0 + W^\epsilon_{jk} + (W^\epsilon - W^\epsilon_{jk})][\phi^0 + \epsilon\phi_1 + \epsilon^2\phi_2 + \cdots]$$

$$= (\lambda_j + a_1\epsilon + a_2\epsilon^2 + \cdots)[\phi^0 + \epsilon\phi_1 + \epsilon^2\phi_2 + \cdots]$$

If we apply linearity and the distributive law and use the fact that $H^0(\phi^0) = \lambda_j\phi^0$ we may divide by ϵ and get

$$a_1\phi^0 + \lambda_j\phi_1 + \epsilon(\quad) + \epsilon^2(\quad)$$

$$= H^0(\phi_1) + [W^\epsilon_{jk}/\epsilon](\phi^0) + [(W^\epsilon - W^\epsilon_{jk})/\epsilon](\phi^0) + W^\epsilon\phi_1 + \epsilon(\quad)$$

Projecting on $\mathcal{H}_{\lambda_j} \cap \mathcal{H}_k$ and letting ϵ approach 0 then yields

$$a_1\phi^0 + \lambda_j P_{jk}(\phi_1) = P_{jk}H^0(\phi_1) + \lim_{\epsilon \to 0} (W^\epsilon_{jk}/\epsilon)(\phi^0)$$

But H^0 commutes with P_{jk} and $H^0P_{jk}(\phi_1) = \lambda_j P_{jk}(\phi_1)$. Thus $\lim_{\epsilon \to 0} (W^\epsilon_{jk}/\epsilon)(\phi^0) = a_1\phi^0$, that is, ϕ^0 is an eigenvector of $\lim_{\epsilon \to 0}(W^\epsilon_{jk}/\epsilon)$ with eigenvalue a_1. Assuming a complete set of eigenvalues depending analytically on ϵ and reducing for $\epsilon = 0$ to members of $\mathcal{H}_{\lambda_j} \cap \mathcal{H}_k$ we see that to a first approximation the eigenvalues of H^ϵ into which λ_j splits are of the form $\lambda_j + \epsilon a_1^{(r)}$ where $a_1', a_1'', \ldots$ are the eigenvalues of $\lim_{\epsilon \to 0}(W^\epsilon_{jk}/\epsilon)$ and k varies over all values for which $\mathcal{H}_{\lambda_j} \cap \mathcal{H}_k \neq 0$. If there are no accidental degeneracies k will be uniquely determined by j. Needless to say there are non-trivial mathematical problems involved in putting this formal discussion of perturbation theory on a rigorous basis. These problems have given rise to a considerable literature to which we refer the reader for further details. A summary by Rellich of the state of affairs at that time appears in the Proceedings of the 1950 International Congress of Mathematics. For later work one may consult the index of *Mathematical Reviews* under the names of such authors as Kato, Nagy, and Wolf.

Returning to group theory let us relate the breakup of λ_j into $\lambda_j + \epsilon a_1^{(r)}$ to the decomposition of the L^k on restriction to the subgroup K. To simplify our statements we shall consider only the case in which there are no accidental degeneracies. Thus for each j there is a unique $k = k(j)$ such that $\mathcal{H}_{\lambda_j} = \mathcal{H}_{\lambda_j} \cap \mathcal{H}_k$ and $P_j = P_{jk} \neq 0$. Since $U_xP_j = P_jU_x$ for all x we compute at once that for all x in K, $U_xP_jW^\epsilon P_j = P_jW^\epsilon P_jU_x$. Thus $W^\epsilon_{j,k(j)}$ commutes with the restriction of U to $\mathcal{H}_{\lambda_j} = \mathcal{H}_{k(j)}$ and K. Now U restricted to $\mathcal{H}_{k(j)}$ is just the irreducible representation $L^{k(j)}$ of G and when $L^{k(j)}$ is restricted to K it has a decomposition $L^{k(j)} = \sum_i \ell(k(j), i)N_i$, where the N_i are the irreducible representations of K and the multiplicity function $k,i \to \ell(k,i)$ depends only on K and G. Let $\mathcal{H}_{\lambda_j} = \mathcal{H}_{k(j)} = \mathcal{H}^{N_1}_{\lambda_j} \oplus \mathcal{H}^{N_2}_{\lambda_j} \oplus \cdots \oplus \mathcal{H}^{N_t}_{\lambda_j}$ be the unique decomposition of $\mathcal{H}_{\lambda_j}$ corresponding to the decomposition of the restriction of $L^{k(j)}$ to K into primary components. Then $W^\epsilon_{j,k(j)}$ carries $\mathcal{H}^{N_i}_{\lambda_j}$ into itself and the eigenvalues of $W^\epsilon_{j,(k)}$ in $\mathcal{H}^{N_i}_{\lambda_j}$ have multiplicity which is a multiple of the dimension d_i of N_i. If H^ϵ has no accidental degeneracies then $W^\epsilon_{j,k(j)}$ will have just as many distinct eigenvalues as $L^{k(j)}$ has (not necessarily distinct) irreducible components when restricted to K. Each of these will occur with multiplicity equal to the dimension of the corresponding N. Thus the multiple eigenvalue λ_j of H^0 associated with the irreducible representation $L^{k(j)}$ of G splits on passage to H^ϵ into as many values as the restriction of $L^{k(j)}$ to K has irreducible components. In particular we see that we may label the eigenvalues of H^ϵ with a double index k,i. The first index k gives the irreducible representation of G associated with the eigenvalue of H^0 from which the eigenvalue of H^ϵ has been split and the second index i gives the irreducible representation of K to which the eigenvalue of H^ϵ belongs in its own right. The index i applies only as long as ϵ is small enough so that the different split-apart eigenvalues may be traced back to their origins. When K is commutative and the $\ell(k,i)$ are all 1 or 0, then there will be at most a one-dimensional family of eigenvectors for a given k,i and eigenvalue, and a stationary state of the system will be completely described by giving the energy and the two indices k and i.

3-4 Spherical Symmetry and Spin

Consider the canonical quantization of the classical dynamical system defined by a particle moving in a "central force field." The classical Hamiltonian is $(1/2m)(p_x^2 + p_y^2 + p_z^2) + V(\sqrt{x^2 + y^2 + z^2})$, where V is the real-valued function giving the potential energy as a function of distance from the "force center" and m is the mass of the particle. When m is the mass of the electron and $V(r) = -e^2/r$ where e is the charge on the electron we get the Rutherford model for a hydrogen atom (with neglect of the motion of the much heavier nucleus). The corresponding dynamical operator is a self-adjoint extension H of the formal differential operator

$$\psi \to -(\hbar^2/2m)\left(\frac{\partial^2\psi}{\partial x^2} + \frac{\partial^2\psi}{\partial y^2} + \frac{\partial^2\psi}{\partial z^2}\right) + V(\sqrt{x^2 + y^2 + z^2})\,\psi$$

acting in the Hilbert space $\mathcal{L}^2(E^3)$. Let G denote the group of all rotations about 0,0,0 in E^3. For each such rotation α let $(U_\alpha f)(x,y,z) = f(\alpha^{-1}(x,y,z))$. Then $\alpha \to U_\alpha$ will be a representation of G by unitary operators in $\mathcal{L}^2(E^3)$ and U_α will commute with H for all α. Thus the considerations of the preceding section may be applied and we have a natural breakdown of $\mathcal{L}^2(E^3)$ as a direct sum $\mathcal{H}_1 \oplus \mathcal{H}_2 \oplus \cdots$, where each $\mathcal{H}_j$ is invariant under H and the U_α, and U restricted to $\mathcal{H}_j$ is primary. We shall suppose the notation chosen so that U restricted to $\mathcal{H}_j$ is a direct sum of replicas of the $2j+1$ dimensional representation D_j of G described in Sec. 3-2. It may be shown that each D_j occurs in the reduction of U and occurs with multiplicity ∞. In fact we may regard E^3 as $S \times R^+$, where S is the surface of the unit sphere in E^3 and R^+ is the positive real axis. Thus $\mathcal{L}^2(E^3)$ is naturally isomorphic to $\mathcal{L}^2(S) \times \mathcal{L}^2(R^+)$ and U_α takes the form $U'_\alpha \times I_\alpha$, where $\alpha \to I_\alpha$ is the identity representation in $\mathcal{L}^2(R^+)$ and $\alpha \to U'_\alpha$ is the representation in $\mathcal{L}^2(S)$ induced by the action of the rotation group on S. Now, the rotation group acts transitively on S and the subgroup leaving the point 0,0,1 fixed is the group K_z of all rotations about the z axis. It follows from a general theorem in the theory of group representations that U' contains each D_j just as many times as the restriction of D_j to K_z contains the identity, that is, once. Thus $U' = D_1 \oplus D_2 \oplus D_3 \oplus \cdots$. Let $\mathcal{H}'_j$ denote the unique subspace of $\mathcal{L}^2(S)$ on which U' reduces to D_j. Then $\mathcal{H}_j = \mathcal{H}'_j \otimes \mathcal{L}^2(R^+)$ and U contains D_j infinitely many times.

Consider H restricted to $\mathcal{H}_j$. From the fact that it commutes with all U_α we deduce easily that this restriction H_j has the form $I \times H'_j$ where H'_j is a self-adjoint operator in $\mathcal{L}^2(R^+)$. Straightforward calculation shows that H'_j is a self-adjoint extension of the second-order differential operator:

$$\psi \to -(\hbar^2/2m)\left(\frac{1}{r^2}\frac{d}{dr}(r^2\frac{d\psi}{dr}) - j(j+1)/r^2\right) + V(r)\psi$$

For the choices of V in which we shall be interested this operator has for each j both a continuous spectrum and a point spectrum and has *no* multiplicities. We shall be interested only in the point spectrum. Let $\mathcal{H}''_j$ denote the discrete part of $\mathcal{L}^2(R^+)$ under H'_j. Then $\mathcal{H}'_j$ is a direct sum of one-dimensional subspaces in each of which H'_j is a constant and $\mathcal{H}^0_j = \mathcal{H}^1_j \otimes \mathcal{H}''_j$ = discrete part of $\mathcal{H}_j$ under H_j, is a direct sum of $2j+1$ dimensional subspaces in each of which H_j is a constant and U reduces to D_j. For the particular case in which $V(r) = -a^2/r$ the eigenvalues of H_j are of the form $-(ma^4/2\hbar^2)(k+j+1)^{-2}$ where $k = 0,1,2,\ldots$. Here the k-th eigenvalue of H_j is the same as the $(k+j-j')$-th eigenvalue of H'_j and we do have accidental degeneracy. Each eigenvalue is associated with a number of different D_j's and has multiplicity equal to the sum of their dimensions. It is customary to call $(k+j+1) = n$ the total or *principal* quantum number of all eigenvectors in the $0+1+3+\cdots+(2n-1)+1 = n^2$ dimensional space of all eigenvectors with $-(ma^4/2\hbar^2n^2)$ as eigenvalue. Those eigenvectors lying in $\mathcal{H}_j$ are said to have *azimuthal* quantum number j. We shall only consider cases in which $V(r)$ is sufficiently close to $-a^2/r$ for some a so that the eigenvalues will be given by $-a^4/2\hbar^2(k+j+1+x_{j,k})^2$, where $x_{j,k}$ is between 0 and $\frac{1}{2}$ and depends chiefly on j—tending to 0 with large j. In these cases the eigenvalues associated with a particular j may be arranged in increasing order and by analogy with the above we shall say that the eigenvectors of H_j whose eigenvalue is the k-th member of this series have principal quantum number $k+j$ and azimuthal quantum number j. The set of all eigenvectors of H having principal quantum number n and azimuthal quantum number j is then a $2j+1$ dimensional subspace of $\mathcal{L}^2(E^3)$.

The principal quantum number of an eigenvector of H determines the (approximate) energy of the corresponding stationary state. What about the azimuthal quantum number? Does it too define a value for some observable? The answer is yes. Recall that the angular momentum observables are $\hbar$ times the infinitesimal generators of the one-parameter groups of rotations around the various axes. It follows from this that each angular momentum observable carries each $\mathcal{H}_j$ into itself and commutes with H. Moreover the square of the *total angular momentum* observable, that is, the sum of the squares of the angular momentum observables in three mutually perpendicular directions, commutes with the U_α and hence is a constant in each irreducible subspace of $\mathcal{H}_j$. This constant can depend only upon j and calculation shows that it is $j(j+1)\hbar^2$. Thus a state of principal quantum number n and azimuthal quantum number j is one in which both the energy and the total angular momentum have definite values with probability 1, the latter being $\hbar\sqrt{j(j+1)}$.

The degeneracy caused by spherical symmetry can be removed by removing the symmetry. In fact it is enough to remove part of the symmetry. In dealing with an atom this is most conveniently done by putting the atom in a uniform magnetic field so that the symmetry group is the subgroup K of G consisting of all rotations about the line through the origin parallel to the direction of the field. Since K is Abelian it has only one-dimensional irreducible representations and the corresponding symmetry produces no degeneracies. Applying the theory of the preceding section and recalling that D_j restricted to K break up into $2j+1$ one-dimensional representations we expect each eigenvalue associated with a state of azimuthal quantum number j to break up into $2j+1$ different values when the spherical symmetry is reduced to cylindrical symmetry. The splitting apart of the eigenvalues —and hence of spectral lines—in a magnetic field is called the *Zeeman effect*.

If we choose a particular axis, say the z axis, then the restriction of U to K_z and the $2j+1$ dimensional subspace of all eigenvectors with principal quantum number n and azimuthal quantum number j breaks this $2j+1$ dimensional subspace up into $2j+1$ one-dimensional invariant subspaces. The vectors in the one-dimensional subspace in which U restricted to K becomes $\theta \to e^{im\theta}I$ are said to have *magnetic* quantum number m. There is just one stationary state with given values for all three quantum numbers and the three quantum numbers index a basis of eigenvectors for the discrete part of H. Of course the three quantum numbers cannot vary independently. We have $|m| \leq j$, $j = 0,1,2,\ldots,n-1$, $n = 1,2,3,\ldots$. The magnetic quantum number m also has a direct physical significance. One computes easily that a state with magnetic quantum number m is one in which the z component of the angular momentum is $\hbar m$ with probability 1.

It is important to note that magnetic quantum numbers always refer to a particular direction in space. The angular momentum operators about different axes do not commute and a state of magnetic quantum number m, in which the angular momentum about some axis has the sharp value $\hbar m$, will be a state in which the angular momentum about other axes will be dispersed and will take on all possible values with non-zero probabilities.

The above analysis applies not only to the hydrogen atom but to a number of other atoms, such as lithium, sodium, and potassium, in which there is a single electron that plays the central role and that can be assumed to move in a spherically symmetrical field. However, when one examines the spectra of these elements closely one finds that for each energy eigenvalue (with $j \neq 0$) predicted by the above theory there are, in fact, two values very close together. Moreover in a magnetic field the two components of an eigenvalue which should split into $2j+1$ values split instead into $2j$ and $2j+2$ values, respectively. This effect in hydrogen is masked by a relativity effect that just compensates for it.

It is not difficult to modify the theory so as to account for the phenomena just described. If we replace $\mathcal{L}^2(E^3)$ by $\mathcal{L}^2(E^3) \times \mathcal{H}_0$, where $\mathcal{H}_0$ is some Hilbert space of finite dimension n, and replace each self-adjoint operator Q in $\mathcal{L}^2(E^3)$ by $Q \times I$ we shall have a quantum system just like the one we started with except for having all eigenvalue multiplicities multiplied by n and for having certain new observables. This resulting degeneracy in $H \times I$ can be attributed to the fact that $H \times I$ commutes with all $I \times V$, where V is a unitary operator in $\mathcal{H}_0$. If it is removed by adding a "small perturbation" to $H \times I$ each eigenvalue of $H \times I$ will split into n parts. To obtain the observed splitting into two parts we need only take $n = 2$. How can such a shift from $\mathcal{L}^2(E^3)$ to $\mathcal{L}^2(E^3) \times \mathcal{H}_0$ be reconciled with the considerations that led us to the canonical quantization? Looking back we see that this shift is inconsistent only with our frankly tentative assumption that the Q_j form a *complete* commuting family. Moreover the discussion on p. 87 suggests that passage from $\mathcal{L}^2(E^3)$ to $\mathcal{L}^2(E^3) \times \mathcal{H}_0$ is the only way of dropping this assumption without losing the Heisenberg commutation relations.

Accepting $\mathcal{L}^2(E^3) \times \mathcal{H}_0$, where $\mathcal{H}_0$ is two dimensional, as the Hilbert space of our system we have a number of questions to answer. (1) What term is to be added to $H \times I$ in order to get the dynamical operator of the modified system? (2) What is the physical significance of the observables attached to self-adjoint operators of the form $I \times T$ where T is a self-adjoint operator in $\mathcal{H}_0$? (3) How must our discussion of rotational symmetry and quantum numbers be modified so as to be consistent with the changed state space and explain the replacement of $2j+1$ by $2j$ and $2j+2$?

It will be convenient to answer these questions in reverse order and to begin by giving some brief indications concerning a generalization of the theory of group representations. A group representation U always defines a homomorphism of the group into the group of all automorphisms of the orthocomplemented lattice of closed subspaces of the Hilbert space $\mathcal{H}(U)$. However the converse need not be true even if one makes the appropriate continuity assumptions. Although each individual automorphism may be implemented by a unitary operator this operator is unique only up to a multiplicative constant and it may not be possible to make the choice so that $U_{xy} = U_x U_y$. On the other hand, no matter how it is made we have $U_{xy} = \sigma(x, y) U_x U_y$ where σ is some function from $G \times G$ to the complex numbers of modulus 1. $x \to U_x$ is then said to be a σ representation or a *projective representation with multiplier* σ. If we replace U_x by $\rho(x)U_x = U'_x$ then $x \to U'_x$ is a σ' representation where

$$\sigma'(x,y) = \sigma(x,y)\rho(xy)/\rho(x)\,\rho(y)$$

The σ and σ' are said to be similar multipliers. It is clear that the similarity classes of multipliers form a group under pointwise multiplication. When this group (called the multiplier group) reduces to the identity we need only consider ordinary representations. As indicated on p. 82 this is the case when G is the real line. When G is the rotation group in E^3, however, the multiplier group contains one non-trivial element and we are forced to consider projective representations. It is quite easy to describe all of them. To this end let SU(2) denote the group of all 2×2 unitary matrices of determinant 1 and let N denote the two-element center consisting of $\begin{pmatrix}1&0\\0&1\end{pmatrix}$ and $\begin{pmatrix}-1&0\\0&-1\end{pmatrix}$. It is not difficult to show that SU(2)/N is isomorphic to the rotation group. Thus the irreducible representations D_j for $j = 1,2,3,\ldots$ define irreducible representations D'_j of SU(2) which reduce to the identity in N. Let L be any irreducible representation of SU(2). Then, by Schur's lemma, L restricted to N must be a constant that is clearly 1 or -1. If it is 1 for both elements of N then L defines an irreducible representation of SU(2)/N and hence must be one of the D'_j. If it is -1 then each member of SU(2)/N has two unitary operators assigned to it, each of which is the negative of the other. Making the choice in an "arbitrary" (but measurable) fashion one obtains a projective representation of SU(2)/N whose multiplier is not trivial. It turns out that SU(2) has one and only one such irreducible representation of every even dimension. We denote the one of dimension 2m by $D'_{m-1/2}$. Thus D'_j is defined for = 1, 3/2, 2, 5/2, 3, ... and always has dimension 2j + 1. When j is an integer D'_j defines the representation D_j of SU(2)/N. When j is a half-integer D'_j defines a projective representation of SU(2)/N which we also denote by D_j. In this way we get all projective representations of SU(2)/N. It turns out that the Clebsch-Gordan series

$$D_j \otimes D'_j = D_{|j-j'|} \oplus D_{|j-j'|+1} \oplus \cdots \oplus D_{j+j'}$$

is valid whether j is an integer or half-integer; direct sums and tensor products being defined in the obvious fashion for projective representations. The two-dimensional representation $D'_{1/2}$ is especially easy to describe—it is just the representation carrying each member of SU(2) into itself.

Returning to our questions we must replace the representation $\alpha \to U_\alpha$ by some representation having $\mathcal{L}^2(E^3) \times \mathcal{H}_0$ as its space. A deeper analysis than we shall give shows that it must be of the form

$U \otimes L$, where L is some two-dimensional ordinary or projective representation of our group. Now the rotation group has just two such: $D_{1/2}$ and $D_0 \oplus D_0$. It follows from the Clebsch-Gordan formula and our earlier analysis that the irreducible subrepresentations of $U \otimes L$ are $D_{j-1/2}$ and $D_{j+1/2}$ in the first case and D_j in the second, where $j = 0,1,2,3,\ldots$. Only the first possibility is consistent with the experimental fact that the energy levels break into $2j$ and $2j+2$ parts instead of $2j+1$ parts. Thus we are led to suppose that the automorphism of our system produced by the rotation α in space is implemented by the unitary operator $U_\alpha \times D_{1/2\alpha}$ and this assumption explains the observed splitting of the energy levels.

The replacement of U by $U \otimes D_{1/2}$ has an effect on the angular momentum observables. The restriction of $U \otimes D_{1/2}$ to the one-parameter subgroup K_z of rotations about the z axis has an infinitesimal generator of the form $i[\Omega^z \times I + I \times \Omega_0^z]$, where $\hbar\Omega^z$ is the self-adjoint operator corresponding to the z component of angular momentum in our naive model and Ω_0^z is a self-adjoint operator in the two-dimensional space $\mathcal{H}_0$. In line with our general principles we define the observable corresponding to $\hbar(\Omega^z \times I) + \hbar(I \times \Omega_0^z)$ to be the angular momentum of our system about the z axis. It differs from what one gets in the canonical quantization by the term $\hbar[I \times \Omega_0^z]$ which turns out to have eigenvalues $\pm \frac{1}{2}\hbar$. This observable is called the *spin angular momentum* (about the z axis). The reason for this terminology is that the splitting of spectral lines into pairs had been "explained" in the old quantum theory by postulating an extra degree of freedom for the electron—a spinning about its own axis. Although an electron as now conceived is too far from a classical particle to be thought of as "really spinning," physicists still find the image a helpful one. From our point of view the content of the spin hypothesis lies in the existence of the supplementary angular momentum-like observables associated with the operators $\hbar[I \times \Omega_0^a]$, where $a = x,y,$ or z. For obvious reasons the observable corresponding to the term $h[\Omega^z \times I]$ is called the orbital angular momentum (about the z axis). Of course, only the total angular momentum about an axis will be conserved.

The operators defining the spin angular momenta have the interesting property of commuting with the operators defining the position observables. They do not commute with one another but any one of them added to the operators defining the position observables produces a complete commuting family of observables. Moreover it is clear that any self-adjoint operator in $\mathcal{L}^2(E^3) \otimes \mathcal{H}(D_{1/2})$ is a simple algebraic combination of self-adjoint operators of the form $A \times I$ and those defining the spin angular momenta. Thus the new observables produced by our passage from $\mathcal{L}^2(E^3)$ to $\mathcal{L}^2(E^3) \otimes \mathcal{H}(D_{1/2})$ can all be traced to the existence of the spin angular momenta.

The dynamical operator for our modified system may be written in the form $H \times I + J$, where H is the dynamical operator coming from the canonical quantization and J is the so-called "spin perturbation." J can be expressed simply and elegantly in terms of the classical potential V and the angular momentum observables as follows:

$$J = \frac{1}{2m^2c^2} \frac{V'(\sqrt{x^2 + y^2 + z^2})}{\sqrt{x^2 + y^2 + z^2}} (\Omega^x \times \Omega_0^x + \Omega^y \times \Omega_0^y + \Omega^z \times \Omega_0^z)$$

where c is the velocity of light. We shall not attempt to justify this formula on theoretical grounds. As the occurrence of c suggests, to do so would involve a discussion of the special theory of relativity and take us beyond the scope of this course. Finally let us look at the effect of the existence of the spin angular momenta on our indexing of energy eigenvectors by quantum numbers. The spin perturbation J is small enough to allow us to relate each eigenvector of $H \times I + J$ to one of $H \times I$ and thus we need only index the eigenvectors of $H \times I$. The $2\ell + 1$ dimensional space of all eigenvectors of H with principal quantum number n and azimuthal quantum number ℓ defines a $2(2\ell + 1)$ dimensional subspace of eigenvectors of $H \times I$. By the Clebsch-Gordan formula $U \otimes D_{1/2}$ restricted to the subspace is the direct sum $D_{\ell-1/2} \oplus D_{\ell+1/2}$. Accordingly our space of eigenvectors breaks up as a direct sum of a 2ℓ-dimensional part and a $2\ell + 2$ dimensional part. The eigenvectors in the first of these are said to have principal quantum number n, azimuthal quantum number ℓ, inner quantum number $j = \ell - \frac{1}{2}$, and spin quantum number $s = -\frac{1}{2}$. Those in the second are said to have principal quantum number n, azimuthal quantum number ℓ, inner quantum number $j = \ell + \frac{1}{2}$, and spin quantum number $\frac{1}{2}$. Knowing any two of the three quantum numbers ℓ, j, and s determines the other according to the equation $j + s = \ell$. The quantum number j, unlike ℓ and s, has a precise meaning apart from the assumption that J is small. The principal quantum number n determines the energy eigenvalue only up to the small correction produced by J. The set of all eigenvectors with fixed values for the principal, azimuthal, and spin quantum numbers is $2j+1$ dimensional where j is the inner quantum number. The total angular momentum operator $\hbar[\Omega^z \times I + I \times \Omega_0^z]$ for the z axis has eigenvectors that define a natural basis in this $2j+1$ dimensional space. The eigenvalues vary from $-\hbar j$ to $\hbar j$ in jumps of $\hbar$. Since j is a half-integer all of these eigenvalues are half-integral multiples of h. The half-integer concerned is called the magnetic quantum number m of the corresponding vector and state. There is just one state with given principal, inner, magnetic, and spin quantum numbers n, j, m, s and three can be chosen arbitrarily subject to the requirements: $s = \pm\frac{1}{2}$, $|m| \leq j$ and m integral, $j = \frac{1}{2}, \frac{3}{2}, \frac{5}{2}, \ldots$, n integral and greater than $j + s$.

Because the basic representation of the rotation group which occurs is $U \otimes D_{1/2}$ the electron is said to be a "particle with spin 1/2." For other "elementary particles" other representations can replace $D_{1/2}$. When D_j occurs one says that the particle has spin j. The photon, for example, has spin 1 and there are mesons with spin 0.

3-5 The n-Electron Atom and the Pauli Exclusion Principle

The classical mechanical system representing an n-electron atom according to Rutherford's theory is determined by the Hamiltonian:

$$\sum_{j=1}^{n} \frac{(p_x^j)^2 + (p_y^j)^2 + (p_z^j)^2}{2m} - \sum_{j=1}^{n} \frac{n e^2}{\sqrt{x_j^2 + y_j^2 + z_j^2}} - \sum_{\substack{i,j=1 \\ i \neq j}}^{n} \frac{e^2}{r_{i,j}}$$

provided that we neglect the motion of the nucleus. Here m and e are the mass and charge, respectively, of the electron and $r_{i,j} = \sqrt{(x_i - x_j)^2 + (y_i - y_j)^2 + (z_i - z_j)^2}$. According to the canonical quantization the corresponding quantum system would have $\mathcal{L}^2(E^{3n})$ as its Hilbert space and a self-adjoint extension H of the formal differential operator:

$$\psi \to -\frac{\hbar}{2m} \sum_{j=1}^{n} \left(\frac{\partial^2 \psi}{\partial x_j^2} + \frac{\partial^2 \psi}{\partial y_j^2} + \frac{\partial^2 \psi}{\partial z_j^2} \right) - \frac{ne^2}{\hbar} \sum_{j=1}^{n} \frac{\psi}{\sqrt{x_j^2 + y_j^2 + z_j^2}}$$

$$- \frac{e^2}{\hbar} \sum_{\substack{i,j=1 \\ i \neq j}}^{n} \frac{\psi}{r_{i,j}}$$

Let us write $H = H_0 + H''$, where H_0 is a self-adjoint extension of the formal differential operator obtained by dropping the last two terms. Then if we replace $\mathcal{L}^2(E^{3n})$, as we may, by $\mathcal{L}^2(E^3) \otimes \mathcal{L}^2(E^3) \otimes \cdots \otimes \mathcal{L}^2(E^3)$ we see that

$$H_0 = \sum_{j=1}^{n} I \times I \times \cdots \times \underset{(j\text{-th factor})}{H'} \times I \times \cdots \times I$$

where H' is a self-adjoint extension of the formal differential operator:

$$\psi \to -\frac{\hbar}{2m}\left(\frac{\partial^2 \psi}{\partial x^2} + \frac{\partial^2 \psi}{\partial y^2} + \frac{\partial^2 \psi}{\partial z^2} \right) - \frac{e^2}{\hbar\sqrt{x^2 + y^2 + z^2}} \psi$$

in $\mathcal{L}^2(E^3)$. The term H'' is due to the interactions between the electrons and prevents a complete tensor product decomposition of the system. However it turns out to be possible to approximate it by an operator of the form:

$$\sum_{j=1}^{n} (I \times I \times \cdots \times H'' \times I \times \cdots \times I)$$

where H'' is an operator in $\mathcal{L}^2(E^3)$ of the form $\psi \to f(\sqrt{x^2+y^2+z^2})\psi$. Thus we may write

$$H = \sum_{j=1}^{n} I \times I \times \cdots \times (H' + H'') \times I \times \cdots \times I + P$$

where P is a "small perturbation."

We saw in the last section that the quantization is incorrect even in the one-electron case, and in view of the results of that section we replace it at once by one in which each $\mathcal{L}^2(E^3)$ is replaced by $\mathcal{L}^2(E^3) \otimes \mathcal{H}(D_{1/2})$ and $H' + H''$ by $(H' + H'' \times I) + J$, where J is another small perturbation. Thus the Hilbert space of our system is a 2n-fold tensor product—n factors being infinite dimensional and n being two dimensional. If we ignore both J and P we have a rather large symmetry group. Every permutation of n symbols and every set of n rotations in E^3 defines a unitary operator that commutes with the unperturbed dynamical operator. In addition we have a way of describing stationary states by quantum numbers. Choosing some axis, say the z axis, for reference, let $\phi_1, \phi_2, \ldots, \phi_n$ be a set of eigenstates of $H' + H''$, where ϕ_k has quantum numbers n_k, j_k, m_k, s_k. Then $\phi_1 \times \phi_2 \times \cdots \times \phi_n$ will be an eigenvector of the unperturbed dynamical operator which we can describe by the $4n$ quantum numbers $n_1, n_2, \ldots, n_n, j_1, j_2, \ldots, j_n, m_1, m_2, \ldots, m_n, s_1, s_2, \ldots s_n$. Moreover these eigenvectors will form a basis for the discrete part of the operator. It is supposed that the actual dynamical operator is close enough to the unperturbed or symmetrized one so that each actual eigenvector can be associated with a unique $\phi_1 \times \cdots \times \phi_n$ and hence described by a set of $4n$ numbers. One thinks of a state in which the first electron is in a state with quantum numbers n_1, j_1, m_1, s_1, the second as in a state with quantum numbers n_2, j_2, m_2, s_2, etc.

When one tries to correlate the spectroscopic data with this theory one finds that there are a great many missing eigenvectors. Indeed those and only those occur that lie in the subspace:

$$(\mathcal{L}^2(E^3) \otimes \mathcal{H}(D_{1/2})) \textcircled{A} (\mathcal{L}^2(E^3) \otimes \mathcal{H}(D_{1/2})) \textcircled{A} \cdots \textcircled{A} (\mathcal{L}^2(E^3) \otimes \mathcal{H}(D_{1/2}))$$

of the full n-fold tensor product. This is explained by supposing that the canonical quantization is wrong in another way. The basic Hilbert space of our system is the n-fold anti-symmetric tensor product of $\mathcal{L}^2(E^3) \otimes \mathcal{H}(D_{1/2})$ with itself and not the full tensor product. The dynamical operator is the restriction to the subspace of anti-symmetric tensors of the original dynamical operator.

This drastic reduction of the state space has profound effects on the fundamental observables—eliminating some and identifying others. Indeed, the discussion given on pp. 110 and 111 for symmetric products applies almost word for word. Electrons, like photons, are more like disturbances in an elastic medium than like classical particles, and interchanging two of them makes no physical difference whatever.

In the terminology of p. 111 electrons are fermions. The fact that photons are bosons and have spin 1 and that electrons are fermions and have spin $\frac{1}{2}$ turns out to exemplify a general law. Bosons always have integral spin and fermions always have half-integral spin.

Let $\phi_1, \phi_2, \ldots, \phi_n$ be eigenvectors of $H' + H''$ with quantum numbers $n_1, n_2, \ldots, j_1, j_2, \ldots, m_1, m_2, \ldots, s_1, s_2, \ldots$. Then the eigenvectors $\phi_{i_1} \times \phi_{i_2} \times \cdots \times \phi_{i_n}$ all correspond to the same eigenvalue of the unperturbed dynamical operator in the full tensor product. Here $i_1, i_2, \ldots, i_n$ is an arbitrary permutation of $1, \ldots, n$. Now it is not difficult to show that the set of all linear combinations of the $\phi_{i_1}, \ldots, \phi_{i_n}$ intersects the anti-symmetric tensor product of the $\mathcal{L}^2(E^3) \otimes \mathcal{K}(D_{1/2})$ in a space which is either one or zero dimensional—the intersection being one dimensional if and only if the ϕ_i are linearly independent. It follows that if the quantum number quadruples $n_1\ j_1\ m_1\ s_1$, $n_2\ j_2\ m_2\ s_2, \ldots$ are all distinct then there is just one corresponding stationary state of the system that is independent of the order in which the quadruples occur. Moreover if any two are equal then there is no corresponding state at all. The fact that there is no stationary state in which two or more electrons have all four quantum numbers the same is called the *Pauli exclusion principle*. It was formulated before the discovery of quantum mechanics in terms of the concepts of the old quantum theory.

When the canonical quantization of the Rutherford atom has been corrected as indicated above it provides a mathematical model for atomic phenomena which has had extraordinary success in explaining and interpreting the facts of spectroscopy and chemistry. Indeed it seems, in principle at least, to reduce the whole of chemistry to a set of problems in pure mathematics. Unfortunately these problems are for the most part of such extraordinary complexity that even with the aid of the new computing machines we are far from being able to dispense with the experimental chemist. Moreover the model is known to be only an approximation—though a very good one under ordinary conditions—since it does not take relativity into sufficient account. When energies are high enough so that relativistic effects become important one needs a new and subtle theory which so far is only partially and imperfectly worked out.

In many instances the quantum mechanics of the atom provides a relatively easy qualitative insight into the nature of chemical phenomena even when an exact quantitative theory is out of reach or very difficult. As an example we conclude these lectures with a brief account of the quantum-mechanical explanation of the existence of the periodic table of the elements. The reader will see that the modifications in our model connected with spin and the exclusion principle play a fundamental role. Unlike the facts of spectroscopy, the facts of chemistry would be extremely and qualitatively different if the canonical quantization actually applied to the atom.

If it were not for the Pauli exclusion principle the most stable stationary state of an atom would be one in which every electron is in a state of principal quantum number 1. One would expect a gradual, more or less monotone change in the properties of an atom as the number of electrons increased and no periodic behavior. Because of the Pauli principle, however, there can be at most two electrons in states with principal quantum number 1. When $n = 1$, $\ell = m = 0$, and s can take on only the two values $\pm \frac{1}{2}$. Thus when we come to the three-electron atom, lithium, at least one electron must be in a state with $n = 2$. Altogether there is room for 8 electrons with $n = 2$. We may have $\ell = 0$, $m = 0$, $s = \pm \frac{1}{2}$, and $\ell = 1$, $m = -1, 0, 1$, $s = \pm \frac{1}{2}$; that is, there are two with $\ell = 0$ and six with $\ell = 1$. However when we get to sodium, the atom with eleven electrons, one of them is forced into a state with $n = 3$.

And so it goes. There are in all $2(2.0 + 1) + 2(2.1 + 1) \cdots = 2(1 + 3 + 5 + \cdots) = 2n^2$ different sets of quantum numbers with principal quantum number n. Thus it should not be until we get to the $2 + 8 + 18 + 1 = 29$-electron atom, copper, that an electron is forced to go into a state with $n = 4$. Actually, however, the electron interactions distort the energy eigenvalues to such an extent that a higher principal quantum number does not always imply a higher energy. As a closer analysis shows the actual ordering is as follows where the first integer is n and the second is ℓ: 1-0, 2-0, 2-1, 3-0, 3-1, 4-0, 3-2, 4-1, 5-0, 4-2, 5-1, 6-0, 4-3, 5-2, 6-1, 7-0, 6-2. Moreover, when the energy eigenvalues of two n's are close together the corresponding states may fill up in reverse order. The actual situation is given in Table 1, which indicates how many electrons are in states of each type from the 18-electron atom to the 30-electron one. After zinc the six states of type 4-1 are filled up in order by gallium, germanium, arsenic, selenium, bromine, and krypton. The filling up of the 5-0, 4-2, and 5-1 states proceeds much as 4-0, 3-2, and 4-1 giving another series of 18 beginning with an alkali metal and ending with an inert gas. It is only when we get to a 58-electron atom that an electron goes into a state with $\ell = 3$. There are 14 states of type 4-3 and the fourteen elements whose atoms have their outermost electrons in these states are the so-called *rare earths*. They seem an anomaly because by the time we have an electron ready to go into a 5-3 state the nucleus is too big to be stable.

Table 1
Numbers of Electrons in Various States

Name	1-0	2-0	2-1	3-0	3-1	4-0	3-2	4-1	5-0	Total
Argon	2	2	6	2	6	—	—	—	—	18
Potassium	2	2	6	2	6	1	—	—	—	19
Calcium	2	2	6	2	6	2	—	—	—	20
Scandium	2	2	6	2	6	2	1	—	—	21
Titanium	2	2	6	2	6	2	2	—	—	22
Vanadium	2	2	6	2	6	2	3	—	—	23
Chromium	2	2	6	2	6	1	5	—	—	24
Manganese	2	2	6	2	6	2	5	—	—	25
Iron	2	2	6	2	6	2	5	—	—	26
Cobalt	2	2	6	2	6	2	7	—	—	27
Nickel	2	2	6	2	6	2	8	—	—	28
Copper	2	2	6	2	6	1	10	—	—	29
Zinc	2	2	6	2	6	2	10	—	—	30

The atoms in which the first electron with a new principal quantum number has just been added all have analogous properties. They are hydrogen and the very active alkali metals. Those with one electron less are the inert gases and similarly for other less striking periodicities. The periods are, of course, of varying length. The first is of length 2, the next two of length 8, the next two of length 18, and the last complete one is of length 32. The two members of the first period are similar to the first and last members of the next two. The first two members and the last six members of the periods of eight are similar, respectively, to the first two and last six of the periods of 18. Finally, if the rare earths are omitted the members of the period of 32 that remain are similar to the members of the period of 18.

We have not of course explained why the periodicity in quantum numbers should lead to a periodicity in chemical properties. To do so would require a development of the quantum theory of chemical reactions and take us too far afield.

APPENDIX

1. Superselection rules. In view of the growing interest in the subject it seems desirable to enlarge upon the remarks made in the second paragraph of page 73. We have seen how the classical von Neumann formulation of quantum mechanics is a consequence of seven more or less plausible axioms (I through VI and VIII) and the more or less *ad hoc* axiom VII. The arguments used in the deduction apply more generally, and it is interesting to work out the modified von Neumann formulation to which one is led when one weakens axiom VII so as to allow the logic to have a non-trivial center. For all systems satisfying axioms I through VI, the center of the logic is a σ-complete Boolean algebra, and the simplest assumption one can make about it (after triviality) is that it is atomic or discrete; that is, that every element is a join of minimal elements. But then these minimal elements are unique, and the logic as a whole is uniquely a direct sum of components having trivial centers. Thus a natural slight weakening of axiom VII is: Axiom VII′—the logic $\mathcal{L}$ has a discrete center and each component in the central decomposition is isomorphic to the lattice of all closed subspaces of a separable infinite-dimensional Hilbert space.

Let $\mathcal{L} = \mathcal{L}_1 \oplus \mathcal{L}_2 \oplus \cdots$ be the decomposition in question and let $\mathcal{H}_j$ be the Hilbert space defining $\mathcal{L}_j$. Let $\mathcal{H} = \mathcal{H}_1 \oplus \mathcal{H}_2 \oplus \cdots$. Then $\mathcal{L}$ is isomorphic to a sublattice $\mathcal{L}_0$ of the lattice of all closed subspaces of $\mathcal{H}$; a closed subspace M of $\mathcal{H}$ being in $\mathcal{L}_0$ if and only if M is the closed linear span of subspaces $M_1, M_2, \ldots$, where $M_j \subseteq \mathcal{H}_j$. Let α be any probability measure on $\mathcal{L}$. Let α'_j be its restriction to $\mathcal{L}_j$. Then $\alpha'_j = \gamma_j \alpha_j$, where $\gamma_j \geq 0$, $\Sigma_{j=1}^{\infty} \gamma_j = 1$, and α_j is a probability measure on $\mathcal{L}_j$. Conversely if α_j is a probability measure on $\mathcal{L}_j$ for each $j = 1, 2, \ldots$, and the γ_j are non-negative real numbers such that $\Sigma_{j=1}^{\infty} \gamma_j = 1$, then there exists a unique probability measure α on $\mathcal{L}$ such that $\alpha(M_1, M_2, \ldots) = \gamma_1\alpha_1(M_1) + \gamma_2\alpha_2(M_2) + \cdots$. Axiom XIII, Gleason's theorem, and an obvious adaptation of the argument given on page 75 leads at once to the conclusion that every probability measure on $\mathcal{L}$ defines a state.

Clearly such a measure α cannot define a *pure* state unless $\alpha(\mathcal{H}_j) = 0$ for all j but one. Thus the most general pure state is defined by choosing a unit vector ϕ in some $\mathcal{H}_j$ and setting $\alpha(M_1, M_2, \ldots) = (P_{M_j}(\varphi) \cdot \varphi)$, where P_{M_j} is the projection in M_j. The argument at the bottom of page 75 allows us to identify observables with self-adjoint operators in $\mathcal{H}$ just as before. However not every self-adjoint operator in $\mathcal{H}$ occurs. The self-adjoint operator A will define an observable if and only if all projections P_E^A in the corresponding projection-valued measure are projections on members of $\mathcal{L}_0$. This is the case if and only if A carries each $\mathcal{H}_j$ into itself. In sum we have exactly the situation described on page 76, with pure states corresponding to unit vectors in the Hilbert space $\mathcal{H}$ and observables corresponding to self-adjoint operators. However, it is no longer true that all self-adjoint operators and unit vectors occur. A unit vector ϕ occurs if and only if ϕ lies in one of the closed subspaces $\mathcal{H}_j$, and the self-adjoint operator A occurs if and only if $A(\mathcal{H}_j) \subseteq \mathcal{H}_j$ for all j. In the terminology of Wightman, Wick, and Wigner [*Phys. Rev.*, 101-105, **88** (1952)], a "superselection rule" operates between $\mathcal{H}_j$ and $\mathcal{H}_k$ whenever $j \neq k$. In the paper just cited these authors give physical arguments for supposing that superselection rules must exist. The fact that the existence of superselection rules is equivalent to replacing axiom VII by axiom VII′ has been noted independently by Mr. Henry Mitchell.

An equivalent formulation of axiom VII′ is the following: The logic $\mathcal{L}$ is isomorphic to the lattice of all closed subspaces of a Hilbert space $\mathcal{H}$ whose projections lie in a von Neumann algebra whose center is discrete and which has no finite components. If we remove the restrictions on the von Neumann algebra we get a further weakening of axiom VII in which we no longer demand a discrete center and permit the "infinitesimal" components in the central decomposition to be isomorphic to the set of all projections in a von Neumann-Murray factor of type II or III. We shall not attempt a complete analysis of this more general situation but content ourselves with two remarks: (1) If the center has no discrete part then pure states cannot exist. (2) When the center is not discrete there can exist one-parameter groups of automorphisms which act effectively on the center. The infinitesimal generators of such one parameter groups do *not* correspond to observables.

For a somewhat different discussion of superselection rules and related matters the reader is referred to Jauch, Helv. *Phys. Acta*, **33**, 711-726 (1960), and Jauch and Misra, *ibid.*, **34**, 699-709 (1961).

2. Supplementary bibliographical remarks. The reader will find a detailed systematic account of C_∞ manifolds in Serge Lang, "Introduction to Differentiable Manifolds," Wiley-Interscience, New York,

1962, and in Sigurour Helgasen, "Differential Geometry and Symmetric Spaces," Chap. I, Academic, New York, 1962.

The idea of using information theory to justify the use of the Gibbs canonical ensemble (Section 1-5) occurred independently to the author but seems to have been introduced into the literature by E. T. Jaynes, *Phys. Rev.*, **106**, 620-630 (1957).

The question raised on page 71 about direct integral decomposition has been studied by Arlan Ramsey in his Harvard thesis. It seems to be much less easy to obtain suitably general results than the author supposed when page 71 was written.

The conjecture about simultaneously answerable questions made at the top of page 71 has been proved true by V. S. Varadarajan [*Comm. Pure Appl. Math.*, **15**, 189-217 (1962)]. The proof is imbedded in a rather elegant exposition of "non-commutative probability theory" and its connections with quantum mechanics.

The notion of "intrinsic Hilbert space" utilized in Section 2-6 on generalized coordinates was apparently first published by L. Schwartz in a paper [Publications de l'Institute Statistique, Univ. Paris, **6**, 241-256 (1957)] which also treats intrinsic L^p spaces. The author came upon the notion independently and it appears in the (unpublished) 1956 version of these notes.

Other approaches to quantization in generalized coordinates will be found in the following references:

1. L. Brillouin, "Les tenseurs en mechanique et en elasticité," Chap. IX, Dover, New York, 1946.
2. B. S. de Witt, *Phys. Rev.*, **85**, 653-661 (1952).
3. I. E. Segal, *J. Math. Phys.*, **1**, 468-488 (1960).

The discussion of infinite-dimensional linear systems on pages 106, 107, and 108 was strongly influenced by a paper of J. M. Cook, *Trans. Am. Math. Soc.*, **74**, 222-245 (1953). For further developments and references on the topic see I. E. Segal, "Mathematical problems of relativistic physics," American Mathematical Society, 1963.

A CATALOG OF SELECTED

DOVER BOOKS

IN SCIENCE AND MATHEMATICS

Astronomy

BURNHAM'S CELESTIAL HANDBOOK, Robert Burnham, Jr. Thorough guide to the stars beyond our solar system. Exhaustive treatment. Alphabetical by constellation: Andromeda to Cetus in Vol. 1; Chamaeleon to Orion in Vol. 2; and Pavo to Vulpecula in Vol. 3. Hundreds of illustrations. Index in Vol. 3. 2,000pp. 6⅛ x 9¼.
Vol. I: 0-486-23567-X
Vol. II: 0-486-23568-8
Vol. III: 0-486-23673-0

EXPLORING THE MOON THROUGH BINOCULARS AND SMALL TELESCOPES, Ernest H. Cherrington, Jr. Informative, profusely illustrated guide to locating and identifying craters, rills, seas, mountains, other lunar features. Newly revised and updated with special section of new photos. Over 100 photos and diagrams. 240pp. 8¼ x 11. 0-486-24491-1

THE EXTRATERRESTRIAL LIFE DEBATE, 1750–1900, Michael J. Crowe. First detailed, scholarly study in English of the many ideas that developed from 1750 to 1900 regarding the existence of intelligent extraterrestrial life. Examines ideas of Kant, Herschel, Voltaire, Percival Lowell, many other scientists and thinkers. 16 illustrations. 704pp. 5⅜ x 8½. 0-486-40675-X

THEORIES OF THE WORLD FROM ANTIQUITY TO THE COPERNICAN REVOLUTION, Michael J. Crowe. Newly revised edition of an accessible, enlightening book recreates the change from an earth-centered to a sun-centered conception of the solar system. 242pp. 5⅜ x 8½. 0-486-41444-2

A HISTORY OF ASTRONOMY, A. Pannekoek. Well-balanced, carefully reasoned study covers such topics as Ptolemaic theory, work of Copernicus, Kepler, Newton, Eddington's work on stars, much more. Illustrated. References. 521pp. 5⅜ x 8½.
0-486-65994-1

A COMPLETE MANUAL OF AMATEUR ASTRONOMY: TOOLS AND TECHNIQUES FOR ASTRONOMICAL OBSERVATIONS, P. Clay Sherrod with Thomas L. Koed. Concise, highly readable book discusses: selecting, setting up and maintaining a telescope; amateur studies of the sun; lunar topography and occultations; observations of Mars, Jupiter, Saturn, the minor planets and the stars; an introduction to photoelectric photometry; more. 1981 ed. 124 figures. 25 halftones. 37 tables. 335pp. 6½ x 9¼. 0-486-40675-X

AMATEUR ASTRONOMER'S HANDBOOK, J. B. Sidgwick. Timeless, comprehensive coverage of telescopes, mirrors, lenses, mountings, telescope drives, micrometers, spectroscopes, more. 189 illustrations. 576pp. 5⅝ x 8¼. (Available in U.S. only.)
0-486-24034-7

STARS AND RELATIVITY, Ya. B. Zel'dovich and I. D. Novikov. Vol. 1 of *Relativistic Astrophysics* by famed Russian scientists. General relativity, properties of matter under astrophysical conditions, stars, and stellar systems. Deep physical insights, clear presentation. 1971 edition. References. 544pp. 5⅝ x 8¼. 0-486-69424-0

Chemistry

THE SCEPTICAL CHYMIST: THE CLASSIC 1661 TEXT, Robert Boyle. Boyle defines the term "element," asserting that all natural phenomena can be explained by the motion and organization of primary particles. 1911 ed. viii+232pp. 5⅜ x 8½.
0-486-42825-7

RADIOACTIVE SUBSTANCES, Marie Curie. Here is the celebrated scientist's doctoral thesis, the prelude to her receipt of the 1903 Nobel Prize. Curie discusses establishing atomic character of radioactivity found in compounds of uranium and thorium; extraction from pitchblende of polonium and radium; isolation of pure radium chloride; determination of atomic weight of radium; plus electric, photographic, luminous, heat, color effects of radioactivity. ii+94pp. 5⅜ x 8½. 0-486-42550-9

CHEMICAL MAGIC, Leonard A. Ford. Second Edition, Revised by E. Winston Grundmeier. Over 100 unusual stunts demonstrating cold fire, dust explosions, much more. Text explains scientific principles and stresses safety precautions. 128pp. 5⅜ x 8½. 0-486-67628-5

THE DEVELOPMENT OF MODERN CHEMISTRY, Aaron J. Ihde. Authoritative history of chemistry from ancient Greek theory to 20th-century innovation. Covers major chemists and their discoveries. 209 illustrations. 14 tables. Bibliographies. Indices. Appendices. 851pp. 5⅜ x 8½. 0-486-64235-6

CATALYSIS IN CHEMISTRY AND ENZYMOLOGY, William P. Jencks. Exceptionally clear coverage of mechanisms for catalysis, forces in aqueous solution, carbonyl- and acyl-group reactions, practical kinetics, more. 864pp. 5⅜ x 8½.
0-486-65460-5

ELEMENTS OF CHEMISTRY, Antoine Lavoisier. Monumental classic by founder of modern chemistry in remarkable reprint of rare 1790 Kerr translation. A must for every student of chemistry or the history of science. 539pp. 5⅜ x 8½. 0-486-64624-6

THE HISTORICAL BACKGROUND OF CHEMISTRY, Henry M. Leicester. Evolution of ideas, not individual biography. Concentrates on formulation of a coherent set of chemical laws. 260pp. 5⅜ x 8½. 0-486-61053-5

A SHORT HISTORY OF CHEMISTRY, J. R. Partington. Classic exposition explores origins of chemistry, alchemy, early medical chemistry, nature of atmosphere, theory of valency, laws and structure of atomic theory, much more. 428pp. 5⅜ x 8½. (Available in U.S. only.) 0-486-65977-1

GENERAL CHEMISTRY, Linus Pauling. Revised 3rd edition of classic first-year text by Nobel laureate. Atomic and molecular structure, quantum mechanics, statistical mechanics, thermodynamics correlated with descriptive chemistry. Problems. 992pp. 5⅜ x 8½. 0-486-65622-5

FROM ALCHEMY TO CHEMISTRY, John Read. Broad, humanistic treatment focuses on great figures of chemistry and ideas that revolutionized the science. 50 illustrations. 240pp. 5⅜ x 8½. 0-486-28690-8

Math–Geometry and Topology

ELEMENTARY CONCEPTS OF TOPOLOGY, Paul Alexandroff. Elegant, intuitive approach to topology from set-theoretic topology to Betti groups; how concepts of topology are useful in math and physics. 25 figures. 57pp. 5⅜ x 8½. 0-486-60747-X

COMBINATORIAL TOPOLOGY, P. S. Alexandrov. Clearly written, well-organized, three-part text begins by dealing with certain classic problems without using the formal techniques of homology theory and advances to the central concept, the Betti groups. Numerous detailed examples. 654pp. 5⅜ x 8½. 0-486-40179-0

EXPERIMENTS IN TOPOLOGY, Stephen Barr. Classic, lively explanation of one of the byways of mathematics. Klein bottles, Moebius strips, projective planes, map coloring, problem of the Koenigsberg bridges, much more, described with clarity and wit. 43 figures. 210pp. 5⅜ x 8½. 0-486-25933-1

THE GEOMETRY OF RENÉ DESCARTES, René Descartes. The great work founded analytical geometry. Original French text, Descartes's own diagrams, together with definitive Smith-Latham translation. 244pp. 5⅜ x 8½. 0-486-60068-8

EUCLIDEAN GEOMETRY AND TRANSFORMATIONS, Clayton W. Dodge. This introduction to Euclidean geometry emphasizes transformations, particularly isometries and similarities. Suitable for undergraduate courses, it includes numerous examples, many with detailed answers. 1972 ed. viii+296pp. 6⅛ x 9¼. 0-486-43476-1

PRACTICAL CONIC SECTIONS: THE GEOMETRIC PROPERTIES OF ELLIPSES, PARABOLAS AND HYPERBOLAS, J. W. Downs. This text shows how to create ellipses, parabolas, and hyperbolas. It also presents historical background on their ancient origins and describes the reflective properties and roles of curves in design applications. 1993 ed. 98 figures. xii+100pp. 6½ x 9¼. 0-486-42876-1

THE THIRTEEN BOOKS OF EUCLID'S ELEMENTS, translated with introduction and commentary by Sir Thomas L. Heath. Definitive edition. Textual and linguistic notes, mathematical analysis. 2,500 years of critical commentary. Unabridged. 1,414pp. 5⅜ x 8½. Three-vol. set.
Vol. I: 0-486-60088-2 Vol. II: 0-486-60089-0 Vol. III: 0-486-60090-4

SPACE AND GEOMETRY: IN THE LIGHT OF PHYSIOLOGICAL, PSYCHOLOGICAL AND PHYSICAL INQUIRY, Ernst Mach. Three essays by an eminent philosopher and scientist explore the nature, origin, and development of our concepts of space, with a distinctness and precision suitable for undergraduate students and other readers. 1906 ed. vi+148pp. 5⅜ x 8½. 0-486-43909-7

GEOMETRY OF COMPLEX NUMBERS, Hans Schwerdtfeger. Illuminating, widely praised book on analytic geometry of circles, the Moebius transformation, and two-dimensional non-Euclidean geometries. 200pp. 5⅝ x 8¼. 0-486-63830-8

DIFFERENTIAL GEOMETRY, Heinrich W. Guggenheimer. Local differential geometry as an application of advanced calculus and linear algebra. Curvature, transformation groups, surfaces, more. Exercises. 62 figures. 378pp. 5⅜ x 8½. 0-486-63433-7

Physics

OPTICAL RESONANCE AND TWO-LEVEL ATOMS, L. Allen and J. H. Eberly. Clear, comprehensive introduction to basic principles behind all quantum optical resonance phenomena. 53 illustrations. Preface. Index. 256pp. 5⅜ x 8½. 0-486-65533-4

QUANTUM THEORY, David Bohm. This advanced undergraduate-level text presents the quantum theory in terms of qualitative and imaginative concepts, followed by specific applications worked out in mathematical detail. Preface. Index. 655pp. 5⅜ x 8½. 0-486-65969-0

ATOMIC PHYSICS (8th EDITION), Max Born. Nobel laureate's lucid treatment of kinetic theory of gases, elementary particles, nuclear atom, wave-corpuscles, atomic structure and spectral lines, much more. Over 40 appendices, bibliography. 495pp. 5⅜ x 8½. 0-486-65984-4

A SOPHISTICATE'S PRIMER OF RELATIVITY, P. W. Bridgman. Geared toward readers already acquainted with special relativity, this book transcends the view of theory as a working tool to answer natural questions: What is a frame of reference? What is a "law of nature"? What is the role of the "observer"? Extensive treatment, written in terms accessible to those without a scientific background. 1983 ed. xlviii+172pp. 5⅜ x 8½. 0-486-42549-5

AN INTRODUCTION TO HAMILTONIAN OPTICS, H. A. Buchdahl. Detailed account of the Hamiltonian treatment of aberration theory in geometrical optics. Many classes of optical systems defined in terms of the symmetries they possess. Problems with detailed solutions. 1970 edition. xv + 360pp. 5⅜ x 8½. 0-486-67597-1

PRIMER OF QUANTUM MECHANICS, Marvin Chester. Introductory text examines the classical quantum bead on a track: its state and representations; operator eigenvalues; harmonic oscillator and bound bead in a symmetric force field; and bead in a spherical shell. Other topics include spin, matrices, and the structure of quantum mechanics; the simplest atom; indistinguishable particles; and stationary-state perturbation theory. 1992 ed. xiv+314pp. 6⅛ x 9¼. 0-486-42878-8

LECTURES ON QUANTUM MECHANICS, Paul A. M. Dirac. Four concise, brilliant lectures on mathematical methods in quantum mechanics from Nobel Prize-winning quantum pioneer build on idea of visualizing quantum theory through the use of classical mechanics. 96pp. 5⅜ x 8½. 0-486-41713-1

THIRTY YEARS THAT SHOOK PHYSICS: THE STORY OF QUANTUM THEORY, George Gamow. Lucid, accessible introduction to influential theory of energy and matter. Careful explanations of Dirac's anti-particles, Bohr's model of the atom, much more. 12 plates. Numerous drawings. 240pp. 5⅜ x 8½. 0-486-24895-X

ELECTRONIC STRUCTURE AND THE PROPERTIES OF SOLIDS: THE PHYSICS OF THE CHEMICAL BOND, Walter A. Harrison. Innovative text offers basic understanding of the electronic structure of covalent and ionic solids, simple metals, transition metals and their compounds. Problems. 1980 edition. 582pp. 6⅛ x 9¼. 0-486-66021-4

HYDRODYNAMIC AND HYDROMAGNETIC STABILITY, S. Chandrasekhar. Lucid examination of the Rayleigh-Benard problem; clear coverage of the theory of instabilities causing convection. 704pp. 5⅜ x 8¼. 0-486-64071-X

INVESTIGATIONS ON THE THEORY OF THE BROWNIAN MOVEMENT, Albert Einstein. Five papers (1905–8) investigating dynamics of Brownian motion and evolving elementary theory. Notes by R. Fürth. 122pp. 5⅜ x 8½. 0-486-60304-0

THE PHYSICS OF WAVES, William C. Elmore and Mark A. Heald. Unique overview of classical wave theory. Acoustics, optics, electromagnetic radiation, more. Ideal as classroom text or for self-study. Problems. 477pp. 5⅜ x 8½. 0-486-64926-1

GRAVITY, George Gamow. Distinguished physicist and teacher takes reader-friendly look at three scientists whose work unlocked many of the mysteries behind the laws of physics: Galileo, Newton, and Einstein. Most of the book focuses on Newton's ideas, with a concluding chapter on post-Einsteinian speculations concerning the relationship between gravity and other physical phenomena. 160pp. 5⅜ x 8½. 0-486-42563-0

PHYSICAL PRINCIPLES OF THE QUANTUM THEORY, Werner Heisenberg. Nobel Laureate discusses quantum theory, uncertainty, wave mechanics, work of Dirac, Schroedinger, Compton, Wilson, Einstein, etc. 184pp. 5⅜ x 8½. 0-486-60113-7

ATOMIC SPECTRA AND ATOMIC STRUCTURE, Gerhard Herzberg. One of best introductions; especially for specialist in other fields. Treatment is physical rather than mathematical. 80 illustrations. 257pp. 5⅜ x 8½. 0-486-60115-3

AN INTRODUCTION TO STATISTICAL THERMODYNAMICS, Terrell L. Hill. Excellent basic text offers wide-ranging coverage of quantum statistical mechanics, systems of interacting molecules, quantum statistics, more. 523pp. 5⅜ x 8½. 0-486-65242-4

THEORETICAL PHYSICS, Georg Joos, with Ira M. Freeman. Classic overview covers essential math, mechanics, electromagnetic theory, thermodynamics, quantum mechanics, nuclear physics, other topics. First paperback edition. xxiii + 885pp. 5⅜ x 8½. 0-486-65227-0

PROBLEMS AND SOLUTIONS IN QUANTUM CHEMISTRY AND PHYSICS, Charles S. Johnson, Jr. and Lee G. Pedersen. Unusually varied problems, detailed solutions in coverage of quantum mechanics, wave mechanics, angular momentum, molecular spectroscopy, more. 280 problems plus 139 supplementary exercises. 430pp. 6½ x 9¼. 0-486-65236-X

THEORETICAL SOLID STATE PHYSICS, Vol. 1: Perfect Lattices in Equilibrium; Vol. II: Non-Equilibrium and Disorder, William Jones and Norman H. March. Monumental reference work covers fundamental theory of equilibrium properties of perfect crystalline solids, non-equilibrium properties, defects and disordered systems. Appendices. Problems. Preface. Diagrams. Index. Bibliography. Total of 1,301pp. 5⅜ x 8½. Two volumes. Vol. I: 0-486-65015-4 Vol. II: 0-486-65016-2

WHAT IS RELATIVITY? L. D. Landau and G. B. Rumer. Written by a Nobel Prize physicist and his distinguished colleague, this compelling book explains the special theory of relativity to readers with no scientific background, using such familiar objects as trains, rulers, and clocks. 1960 ed. vi+72pp. 5⅜ x 8½. 0-486-42806-0

CATALOG OF DOVER BOOKS